The Stories of the All Father of Norse Mythology

By

W. Wagner

Copyright © 2011 Read Books Ltd.
This book is copyright and may not be
reproduced or copied in any way without
the express permission of the publisher in writing

British Library Cataloguing-in-Publication Data
A catalogue record for this book is available from
the British Library

Contents

Page No.

Odin, Father of the Gods and of the Ases	1
The Sleeping Heroes	13
The Higher Conception of Odin	16
Odin at Geiröd's Palace	18
Odin, the Discovery of the Runes, and God of Poetry and Wisdom	21
Odin's visit to Gunlöd	23
Other Goddesses related to Frigg	37
Holda. Ostara	42
Berchta or Berta	50
Thor, Thunar (Thunder)	56
Thor's Deeds and Journeys	60
Thor's Journey to Utgard	64
Thor's Journey to Thryheim	77
Tyr or Zio	90
Heimdal (Riger)	101
Bragi and Iduna	107

ODIN, FATHER OF THE GODS AND OF THE ASES.

THE prophetess Wöla sat before the entrance of her cave, and thought over the fate of the world. Her prophetic power enabled her to pierce bounds that are impenetrable to the human eye. She saw what was going on near her, what was taking place at a distance. She watched the labours and battles, the patient endurance and the victories of nations and heroes. She saw how Allfather ruled the world, how he kept the giants in submission, how he flung the spear of death over the armies, and afterwards sent his Walkyries to bring to his hall those heroes who had fallen victoriously. Let us now turn our attention to what was revealed to her penetrating sight.

Mother Night was driving in her dark chariot on her accustomed course above Midgard, bringing peaceful slumber to all creatures. The bright boy, Mani (Moon), followed quickly in her steps, and the gloomy mountains were bathed in the light he shed around. Down below in the valley, the maiden, Selke, was wandering beside a stream, which playfully rippled and murmured at the feet of its mistress, and then flowed on quickly, and dashing over the stones that barred its course, flung itself into the depth below. But Selke saw nothing of all this; her eyes were fixed on the fountain from out of which the brook flowed, for there sat a woman wondrously beauteous of countenance, with long shining golden hair, looking down into the clear water in which her form was mirrored. After awhile she rose, and went higher up the steep side of the mountain to the place where grew the healing herbs that the goddess needed for the cure of wounds and sores.

While employed in this peaceful task, the rocky door leading into the interior of the mountain suddenly opened, and a monstrous giant came out from it. No sooner did the fiend sight the lovely maiden than he rushed towards her with a wild yell. She fled, while he pursued her, as higher and higher she climbed, until at length she reached the summit of a lofty rock, which hung over the edge of a great abyss. The hunt-cry from the distance now fell upon her ear, and the baying of hounds, and she knew who was coming to her assistance; but her pursuer drew nearer and nearer, and his icy talons almost grasped her neck; boldly she ventured the tremendous leap—the ground was reached in safety.

The mark of her foot is still to be seen on the rock, and the truth of this assertion can be verified by any one who chooses to go and look at the Maiden's Leap in the Selkethal (Harz Mountains).

The giant saw her take the fearful spring, and, surprised, he hesitated for a moment; but soon regaining courage, he rushed on and took the mighty leap after her. But, like a flash of lightning, and accompanied by loud peals of thunder, a shining spear came flying through the air, and the monster fell with a crash dead into the deep abyss.

The storm rose; it howled through the wood, and Wodan's raging host, the Wild Hunt, rushed past. The great god's nightly following was composed of armed men, armed women and children, hounds and ravens and eagles; and he, the King, preceded them all on horseback; together they stormed over the trembling fields and through the dark quaking forests. Ancient pines were broken down, rocks fell, and the mountains shook to their foundations, for the Father of Victory was on his way to a great battle.

The King had far to go, and his horse had lost a shoe, which forced him to halt for a time. Master Olaf, the smith of Heligoland, was still in his smithy at work in the midnight hour. A storm was howling round the house, and the sea was beating on the shore, when suddenly he heard a loud knocking at his gate.

"Open quick and shoe my horse; I have a long journey to make, and daybreak approaches."

Master Olaf opened the door cautiously, and saw a stately rider standing beside a giant horse. His armour, shield, and helmet were black, a broad sword was hanging at his side, his horse shook its mane, champing the bit and pawing the ground impatiently.

"Whither art thou going at this time of night, and in such haste?" asked the smith.

"I left Norderney yesterday. It is a clear night, and I have no time to lose, as I must be in Norway before daybreak."

"If thou hadst wings, I could believe thee," laughed the smith.

"My horse is swift as the wind. But see, a star pales here and there; so make thee haste, good smith."

Master Olaf tried on the shoe. It was too small, but, lo! it gradually grew and grew, until it had fastened itself round the hoof. The smith was awe-struck, but the rider mounted, and as he did so his sword rattled in its sheath.

"Good-night, Master Olaf," he cried. "Thou hast shod Odin's horse right well, and now I hasten to the battle."

The horse gallopped on over sea and land. A light shone round Odin's head and twelve eagles flew after him swiftly, but could not overtake him. He now began to sing in magic words of the stream of time, and the spirit that works in it, of birth, and of the passage to eternity. And all the time the storm-wind roared, and the waves dashed upon the shore, a harp-like accompaniment to the song. He who has ever heard that music straightway forgets his home and his cravings for the hearth. The sailor on the foaming water, the traveller in the valley and the shady grove, each feels it strangely stirring his soul, each longs to go out at once to Odin.

The warriors were gathered together in the green-wood, armed for the combat; the brave sons of King Eric of the bloody axe, who had lately fallen in battle, were there, and Hakon, too, his brother, the powerful king of Norway. All at once they heard sweet soft sounds in the air, like the sighing of the wind and the whisper of green leaves. Quickly the sounds grew louder, and the storm wind roared through the trees and over the assembled host. "Odin is coming," cried the warriors, "he is choosing his Einheriar." And then the Father of Battles came with his following; he came in the storm that he might rule the combat. He halted high up above the armies in a grey sea of clouds. He called the Walkyries, Gondul and Skogul, before him, and bade them so to lead the chances of the fight, that the bravest should be victorious, and should then be received into the ranks of the Einheriar.

He flung his spear over the contending heroes, and immediately the blast of horns and loud war-cries were heard. A cloud of arrows hissed through the air; javelins and heavy battle-axes broke through helmet and shield; swords were crossed in single combat; blood streamed from innumerable wounds, reddened the armour of the men-at-arms and trickled down upon the flowers that carpeted the crimson ground.

Foremost in the battle was King Hakon fighting with sword and spear. As he cut his way through the enemy's ranks over the fallen men, he heard the Walkyries talking beside him. They were in the midst of the strife, mounted on their white horses, holding their bright shields in front of them, and leaning upon their spears.

"The army of the gods is waxing great," said Gondul, "for the Ases are preparing to welcome Hakon with a goodly train of followers to the glorious home."

The King heard it, and asked: "Is it just that ye should reward me with death, instead of the victory for which I am striving with my might?"

Skogul answered: "We have decreed that thine enemies should give way before thee. Thou shalt win the battle, and then take thy part in the feast of the Einheriar. We will now ride on before thee, and announce that thou art coming to look upon the face of the Father of Victory himself."

When King Hakon ascended to Asgard from the field of glory, Hermodur, the swift, and Bragi, the divine singer, went out to meet him, and said: "Thou shalt have the peace of the Einheriar; receive therefore the draught prepared for the heroes of the Ases." Hereupon the king's helmet and coat of mail were taken off, but he retained his sword and spear, that he might enter the presence of the Father of Victory with his arms in his hands.

This was how the Northern skalds sang of the God of Battles, of the choosers of the dead, and of the fate of heroes. Is it then to be wondered at, that the princes and nobles of those races should have gone forth joyously on their bold Wiking raids, and that they should have esteemed a glorious death on the field of battle far better than to sink to inglorious rest at home? The German bards also sang after this fashion of their heroes; hence the struggle against Rome which lasted four hundred years, and the Germanic raids upon Britain, Gaul, Italy, Spain, and even upon far Africa. The War-god sang his storm-song in their ears; they heard the voices of the Walkyries through the din of the battle; they saw the gates of Walhalla open before them, and the Einheriar signing to them to approach. Therefore the day of battle was in their eyes either a feast of victory, or of entrance into the verdant home of the heroes.

In the foregoing tale, the events of which have been derived from German and Norse sagas and lays, we have seen the chief god of the North as leader of the Wild Hunt, conqueror of the earth-born giant, god of the storm and ruler of the battle; but we must try to get a still deeper insight into his nature.

Wodan, Odin in the North, according to the oldest conceptions.—Wodan was the highest and holiest god of the Germanic races. His name is connected with the German word *Wuth*, and used to be both spelt and pronounced Wuotan, which word did not then mean rage or wrath, as *Wuth* does now, but came from the Old-German *watan*, impf. *wuot*, *i.e.*, to penetrate, to force one's way through anything, to conquer all opposition. The modern German *waten*, and the English *wade*, are derived from the old word, though considerably restricted in meaning. Wuotan was therefore the all-penetrating, all-conquering Spirit of Nature. The Longobards, by a letter-change, called him Gwodan; the Franks, Godan or Gudan; the Saxons, Wode; and the Frisians, Woda. The Scandinavians called him Odin, from which the mythological name Odo was derived. He was known under the names of Muot (courage) and Wold by the South Germans. But everywhere he was regarded as the same great god, and was worshipped as such by the whole Germanic race.

When man had freed himself from the power of the impressions made upon him by nature as a whole, he began to have a more distinct consciousness of certain manifestations of the forces of nature, and after that to pay them divine honours. He then regarded the storm which tore through the forests with irresistible violence, which blew down the cottages of the peasants, and wrecked vessels out at sea, as the ruler of all things, as the god whose anger must be appeased by prayers and sacrifices. At first he was worshipped under the form of a horse or of an eagle, as these were types of strength and swiftness. But when the mastery of the human race over the animal world was better understood, the god was endowed with a human form. He was described in the legends and stories, now as a mighty traveller who studied and tried the dispositions of men, and now as an old man with bald head, or with thick hair and a beard which gained him in the North the name of Hrossharsgrani (horse-hair bearded). He had usually only one eye, for the heavens have but one sun, Wodan's eye. He wore a broad-brimmed hat pulled down low over his forehead, which represented the clouds that encircle the sun, and a blue mantle with golden spangles, *i.e.*, the starry heavens. These attributes again prove him to have been the Spirit of Nature. In the completely developed myth regarding him in the Edda, he was described as being of grand heroic form, with a golden helmet on his head, and wearing a shining breast-plate of chain-mail. His golden ring Draupnir was on his arm, and his spear Gungnir in his right hand. Thus attired, he advanced to attack the Fenris-Wolf, when the Twilight of the Gods was beginning to fall; thus attired, he sat on his throne Hlidskialf, wrapped in the folds of his mantle, and governed gods and men.

There are many tales and traditions about Wodan in his original form of storm-god. They are to be found in Germany, England, France, and Scandinavia, which shows how wide-spread the worship of him was. Chief amongst the stories referring to the old Teutonic god are those of the Wild Hunt, and of the Raging Host.

The Myths of the Wild Hunt and of the Raging Host.— These myths have their origin in the belief that the supreme One takes the souls of the dead to himself, carries them through the air with him, and makes them his followers on his journeys by night. As the Romans regarded Mercury as the leader of the dead, they thought that the Teutons also honoured him as the highest god. The soul was looked upon as aërial, because it was invisible like air. It was held that when a dying man had drawn his last breath, his soul passed out of him into the invisible element. Thus the Hebrews had the same word to express spirit and breath, and the old Caledonians, as Ossian's poems prove, heard the moans and loving words of their dead friends in the whisper of the breeze, in the soft murmur of the waves; they felt that the invisible was near them, when a solitary star sent down its rays to them through the dusk of the evening. The idea of a god has no place in these poems. The Teutons, on the contrary, believed that it was the god himself who bore the spirits of the dead up into his kingdom.

The traditions of the Wœnsjäger, the Wild Huntsman, Wuotan's or the Raging Host, have their origin in heathen times, as their names show, although they have undergone considerable modifications in many respects since then. They arose from the impression made upon the people by phenomena that they could not understand, and which they consequently supposed were caused by some divinity. Every noise sounds strange and mysterious on a quiet night. The solitary traveller passing through forests or over heaths or mountains, when the light of the moon and stars was obscured by drifting clouds, heard the voices of spirits in the hooting of owls, in the creaking of branches, and in the roaring, whistling, and howling of the tempest, and his excited imagination made him think that he saw forms, which became the more distinct the more his superstitious fancy was drawn upon. Forest rangers, solitary dwellers in remote places, especially charcoal-burners, who often spend long stretches of time without seeing a human being, tell strange stories even now-a-days. These tales are founded on the ancient beliefs of the race, are repeated by one man to another, and detached fragments of the old faith are still preserved by tradition.

In Pomerania, Mecklenburg, and Holstein, Wode is said to be out hunting whenever the stormy winds blow through the woods. In Western Hanover it is said to be the Woejäger, in Saterland the Woinjäger, and in other places, the Wild Huntsman that haunts the woods. He is supposed to ride on a white horse, to wear a broad-brimmed hat slouched over his forehead, and a wide cloak (the starry heavens) wrapped round his shoulders. This cloak has gained him the name of Hakel-bärend (Mantel-wearing) in Westphalia. Indeed, the story has even been transferred from the divine to the human.

It was said that Hans von Hakelberg, chief huntsman of the Duke of Brunswick, and an enthusiastic sportsman, liked hunting better than going to church, and used to devote his Sundays as well as week-days to this amusement, for which reason he was condemned to hunt for ever and ever with the storm. His grave is shown near the Klöpperkrug, an inn not far from Goslar, and a picture of both him and his hounds is carved on the headstone of the grave. His burial place is also pointed out in the Söllinger wood, near Uslar.

Wode seldom hunted alone. He was generally surrounded by a large pack of hounds, and accompanied by a number of huntsmen, who all rushed on driven by the storm, shouting and holloaing, in pursuit of a spectral boar or wild horse. He was also said to chase a spectral woman with snow-white breast, whom he could only catch once in seven years, and whom he bound across his saddle when he had at length succeeded in overtaking her. In Southern Germany it was a moss-woman or wood-maiden, a kind of dryad or wood-nymph, whom the Wild Huntsman pursued, and whom he bound to his horse in the same way as the other, when once he had caught her. Perhaps this story represents the autumnal wind blowing the leaves off the trees.

When the people heard the Wild Huntsman approaching them they threw themselves upon their face on the ground, as otherwise they would have been in danger of being carried off by the huntsmen. The story tells us that this was the fate of a ploughman who was caught up by them and taken away to a hot country where black men lived. He did not come home again until many years afterwards. Whoever joined in the holloa of the wild huntsmen was given a stag's leg which became a lump of gold; but whoever imitated the shout jeeringly had a horse's leg thrown to him, which gave out a pestiferous smell and stuck to the scoffer. A little dog was sometimes left on the hearth of a house through which the Wild Huntsman had gone. It immediately began to whine and howl miserably, so as to disturb the whole household. The people had then to get up and brew some beer in egg-shells, whereupon the creature would exclaim: "Although I am as old as the Bohemian Forest, I never saw such a thing in my life before." Then it would jump up, rush off and vanish. But if this charm was not applied, the people of the house were obliged to feed the creature well, and let it lie upon the hearth for a whole year, until Wode returned and took it away with him.

The Wild Hunt generally went on in the sacred season, between Christmas and Twelfth Night. When its shouts were particularly loud and distinct, it was said that it was to be a fruitful year. At the time of the summer solstice, and when day and night become of equal length, the Wild Hunt again passed in the wind and rain, for Wodan was also lord of the rain, and used to ride on his cloud-horse, so that plentiful rains might refresh the earth.

The traditions of the Raging Host much resemble those of the Wild Hunt. They are stories about the army of the dead under the leadership of Wodan. People thought they could distinguish men, women and children as the host passed them at night. Those who had lately died were often seen in it, and sometimes the death of others was foretold by it.

"Walther von Milene!" cried out voices in that terrible army, and Walther, a celebrated warrior, was soon afterwards killed in battle. In this instance the story reminds us of Wish-father, the chooser of the dead, who called the Einheriar to his Walhalla; and still more is this the case, when the Raging Host is described as rushing past like a troop of armed men, when knights and men-at-arms were seen in shining or even fiery armour, and mounted upon black horses, from whose nostrils shot forth sparks of flame. Then it was said that the war-cries of the combatants, the clash of arms and trampling of horses' feet, could be heard above the din of the storm.

Wodan has long since died out of the minds of the people, yet his character and actions are clearly shown in tradition, and his name also appears in proverbial sayings, charms, and invocations. Seventy years ago the Mecklenburg farmers, after the harvest was brought home, used to give their labourers Wodel-beer, a feast at which there was plenty to eat and drink. The people poured out some of the beer upon the harvest field, drank some themselves, and then danced round the last remaining sheaf of corn, swinging their hats and singing:

> "Wôld! Wôld! Wôld!
> hävenhüne weit wat schüt,
> jümm hei dal van häven süt.
> Vulle kruken un sangen hät hei,
> upen holte wässt manigerlei :
> hei is nig barn un wert nig old.
> Wôld! Wôld! Wôld!"*

> "Wold! Wold! Wold!
> The Heaven-Giant knows what happens here;
> From Heaven downwards he does peer.
> He has full pitchers and cans.
> In the wood grows many a thing.
> He ne'er was child, and ne'er grows old,
> Wold! Wold! Wold!"

In Hesse and in Lippe-Schaumburg the harvesters stick a

* Grimm's "Teutonic Mythology," translated by J. S. Stallybrass, vol. i. p. 156. (London: Sonnenschein & Allen.)

bunch of flowers into the last sheaf, and beat their scythes together, exclaiming, Waul; in Steinhude they dance round a bonfire they have lighted on a hill-top, and shout, Waude. In many parts of Bavaria they dance round a straw figure called Oanswald or Oswald (Ase Wodan). But the people have now quite forgotten the Ase and think only of St. Oswald. In these instances the god appears in his highest form as the god of heaven, the giver of good harvests. The Aargau riddle shows him as lord of the starry heavens, who raises the dead up to his bright mansions above:—

> " Der Muot mit dem Breithuot
> Hat mehr Gäste, als der Wald Tannenäste."

> " Muot with the broad hat
> Has more guests than the wood has fir-twigs."

In England the Wild Hunt is called Herlething, from a mythical king Herla, who was once invited by a dwarf to attend his marriage. He followed his entertainer into a mountain, and three hundred years elapsed before he and his attendants returned to the world. Amongst other parting gifts the dwarf gave him a beautiful dog, which the head huntsman was desired to take before him on his horse. At the same time every one was warned not to dismount until the dog jumped down. Several of the king's followers disregarded this, and got down from their horses; but no sooner did they touch the ground than they crumbled away to dust. The dog is still sitting on the saddle bow, and the Wild Hunt is still going on.

In the time of Henry II. it was said to have shown itself in a meadow in full daylight. The blowing of the horns and shouts of the hunters drew the people of the neighbourhood to the place. They recognised some of their dead friends among the huntsmen, but when they spoke to them, the whole train rose in the air, and vanished in the river Wye.

In France, in Wales, and in Scotland, King Arthur is the leader of the Wild Hunt. In France, the Wild Hunt, or Raging Host, is called *Mesnie Hellequin*, the last word of which is evidently derived from Hel (kingdom of the dead), for the leader of the hunt is called the Hel-huntsman. According to other traditions, Charles the Great, Charlemagne, rides in front of the band, while strong Roland carries the banner. We recognise, moreover, the Raging Host (*l'armée furieuse*) under the name of *Chasse de Caïn* (Cain's Hunt), or *Chasse d'Hérode* (Hunt of Herodias, who caused the murder of John the Baptist). Perhaps, however, Hérode really means Hrodso (glory-bearer), one of the names by which Odin was known. Equally famous is *le grand veneur de Fontainebleau*, (the great Huntsman of Fontainebleau), whose shouts were heard beside the royal palace the day before Henry IV. was murdered by Ravaillac. The Raging Host also passed over the heavens twice, darkening the sun, before the Revolution broke out. The populace everywhere believes that its appearance is the foreshadowing of pestilence, or war, or of some other great misfortune.

THE SLEEPING HEROES.

The legend of the Wild Huntsman has, as we have seen from the foregoing, been applied to human beings, and circumstance and place have been added to the tale. There was not always an infernal element clinging to the appearance of the Hunt, for emperors, kings, and celebrated heroes were amongst the representatives of the Father of the Gods. In Lausitz, Dieterbernet— in Altenburg, Berndietrich, the great Ostrogothic king Theoderick of Bern (Verona) was supposed to rush through the air, and vanish in the mountains. In the same way, according to the Northern myth, the Summer Odin, who brought green leaves and flowers, and ripened the golden ears of corn, used to wander away through dark roads in Autumn, and then a false Odin came, and seating himself on the other's throne, sent snowstorms over the wintry earth. Or, as another tale has it, the good god passed the period during which the imposter reigned, sunk in a deep enchanted sleep within a mountain. But no sooner did Spring return, than he rose again in his power, drove the intruder from his throne, and once more scattered his blessings over gods and men.

These conceptions of Allfather, derived from natural phenomena, were so deeply impressed in the mind and very being of the Teutonic race, that they personified them by applying to their early kings and heroes the attributes of Odin. King Henry the Fowler, whose victories over the Slavs, Danes and Hungarians restored the power of the German empire, is supposed to be lying sunk in magic sleep in the Südemer hill near Goslar. Amongst other sleeping heroes is Frederick Barbarossa, the story of whose death in the East is believed by no one, and who was and is still said to lie slumbering in Kyfthäuser.

There are a number of traditions about the ruins of Kyffhäuser and the great Hohenstaufe, who still lives in the memory of his people. The high castle-hill rises sheer above the green fields away over in Thuringia. On its western side, a tower is still in existence. It stands eighty feet high, although with broken walls, and overlooks the wood and piles of stone below. On solemn occasions the emperor is supposed to lead his processions thence, and afterwards to dine there with his followers. According to the legend, the weary old emperor sleeps his "long sleep" in an underground chamber of the castle, with the companions of his travels, Christian of Mayence, Rainald of Cologne, Otto of Wittelsbach, the ancestor of the royal house of Bavaria, and many others besides. Barbarossa's beard has grown round and through the stone table, casks of good old wine, treasures of gold, silver and precious stones are lying about in heaps, and a magic radiance lights up the high vaulted hall; that this is the case is proved by many fortunate eye-witnesses, who at different times have been permitted to enter the room. One of these was a herdsman, who left his cattle browsing amongst the ruins, and went to gather flowers for his sweetheart. He found a strange blue blossom, and no sooner had he put it in his nosegay than his eyes were opened, and he perceived an iron door that he had never seen before. It opened at his touch; he went down a flight of stairs and entered the lighted banqueting hall. There he saw the heroes and their imperial leader sitting round the table, all sound asleep in their chairs.

Barbarossa was awakened by the noise. "Are the ravens still flying round the battlements?" he asked, looking up.

The herdsman said that they were, and the emperor went on: "Then I must sleep for another hundred years."

After that he invited the youth to help himself to as much as he liked of the treasures he saw before him, and not to forget the best.

The herdsman filled his pockets as he was told. When he got out into the open air once more, the door shut behind him with a crash, and he could never find it again, for he had forgotten the best thing, the little blue flower. So the emperor is still sleeping with his heroes in his favourite palace. But the time will come when the empire is in greatest need of him, when the ravens will no longer fly round the battlements; then he will arise in all his might, will break the magic bonds that hold him, and sword in hand fight a great and bloody battle against the enemies of his country upon the Walser Field or on the Rhine. Then he will hang his shield on a withered pear-tree, which will immediately begin to sprout again, and blossom and bear fruit: the glorious old times of the German Empire will return, bringing with them unity and peace in their train.

THE HIGHER CONCEPTION OF WODAN (ODIN).

Wodan, the giver of victory. Ambri and Assi the Winilers, stood fully armed before the warlike Vandals. Their victory or servitude would be decided by the coming battle.

"Give us the victory, Father of Battles," prayed the princes of the Vandals, as they offered up sacrifices to Wodan. And the god answered: "To them will be given the victory who come first before me on the morning of the day of battle."

On the other hand Ibor and Ajo, dukes of the Winilers, went by the counsel of their wise woman, Mother Gambara, into the holy place of Freya, Wodan's wife, and entreated her to aid them.

"Well," said the Queen of Heaven, "let your women go out ere daybreak dressed in armour like the men, their hair combed down over their cheeks and chins, let them take up a position towards the east, and I will give ye a glorious victory."

The dukes did as she commanded.

As soon as the first rosy tints of dawn appeared in the sky, Freya wakened the great Ruler, and pointed eastwards towards the armed host.

"Ha!" said the god in astonishment, "what long-bearded warriors are these?"

"Thou hast named them," answered the queen, "so now do thou give them the victory." And thus the Winilers gained great glory, and were henceforth known by the name of Long Beards Longobards).

As in the Northern myths, the Longobards also held great Wodan to be the giver of victory. But above all other qualities, he was the god who blessed mankind, and brought joy and prosperity to his people.

In the heathen times many games and processions were held in his honour, of which traces still remain in the customs and beliefs of the people. In many districts, for instance, the battle of the false Odin, who usurped the throne for the seven winter months, with the true Odin, who brought blessings and summer into the world, was celebrated by a mimic fight, succeeded by sacrifices and feasting. This lasted for centuries, and was continued until quite recent times in the festivals of the first of May.

A May Count or May King was chosen, and he was generally the best runner or rider, or the bravest in the parish. He was dressed in green and adorned with garlands of may and other flowers. He then hid himself in the wood; the village lads went out to seek him there, and when they had found him, they put him on horseback, and led him with shouts and songs of joy through the village. The May King was allowed to choose a queen to share his honours at the dance and at the feast.

In other places the most modest and diligent of the girls was chosen as Queen of May, and led into the village with the King, which was intended to commemorate the marriage of the Summer Odin with the Earth, whose youth was renewed by the genial Spring. It was at one time a regular practice to have a May-ride in Sweden, at which the May Count, decked in flowers and blossoms, had to fight against Winter, who was wrapped up in furs. May won the victory after a burlesque hand-to-hand engagement.

Odin, the good and beneficent god, was also called Oski, *i.e.*, "wish" in Norse, a word that is related to the German *Wonne* (rapture): he was the source of all joy and rapture.

ODIN AT GEIRÖD'S PALACE.

King Hraudung had two handsome sons, Geiröd and Agnar, the one ten and the other eight years old. The boys one day went out in a boat to fish. But the wind rose to a storm, and carried them far away from the mainland to a lonely islet, where the boat struck and broke in pieces. The boys managed to reach the shore in safety, and found there a cottager and his wife, who took compassion on them and gave them shelter. The woman took great care of the younger brother Agnar throughout the winter, while her husband taught Geiröd the use of arms and gave him much wise counsel. That winter the children both grew wonderfully tall and strong, and this was not surprising, for their guardians had been Odin and his wife Frigg. When spring returned, the boys received a good boat and a favourable wind from their protectors, so that they soon reached their native land. But Geiröd sprang on shore first, shoved the boat out to sea again, and cried, "Sail thou away, Agnar, into the evil spirits' power!" The great waves, as though in obedience to the cruel boy's behest, carried the boat and Agnar far away to other shores. Geiröd hastened joyfully up to the palace, where he found his father on his death-bed. He succeeded to the kingdom, and ruled over all his father's subjects and those he had gained for himself by force of arms and gold.

Odin and Frigg were once sitting on their thrones at Hlidskialf gazing down at the world of mortal men and at their works. "Seest thou," said the Ruler, "how Geiröd, my pupil, has gained royal honours for himself? Agnar has married a giantess in a foreign land, and now that he has returned home, is living in his brother's palace poor and despised." " Still Geiröd is only a base creature, who hoards gold and treats his guests cruelly instead of showing them hospitality," replied the thoughtful goddess. Then Allfather determined to prove his favourite, and to reward him if all were well, but to punish him should he find that the accusation was just. He, therefore, in the guise of a traveller from a far country, started for Geiröd's palace. A broad-brimmed hat, drawn well down over his brows, shaded his face, and a blue cloak was wrapped around his shoulders. But the King had been warned by Frigg of a wicked enchanter, so he had the stranger seized and brought before his judgment-seat.

To all the questions asked him, the prisoner would only reply that his name was Grimnir, and disdained to give further information about himself. Whereupon the king got into a passion, and commanded that the obstinate fellow should be chained to a chair between two fires upon which fresh fuel was to be continually thrown, so that the pain he suffered might induce him to speak out.

The stranger remained there for eight nights, suffering bitter agony, without having had a bite or a sup the whole time, and now the flames were beginning to lick the seam of his mantle. Secretly Agnar, the disinherited, gave him a full horn of beer, which he emptied eagerly to the last drop. Then he began to sing, at first low and softly, but afterwards louder and louder, so that the halls of the castle echoed again, and crowds assembled without to listen to the strain. He sang of the mansions of the blessed gods, of the joys of Walhalla, of the Ash Yggdrasil, of those that dwelt within it, and of its roots in the depths of the worlds.

The halls trembled, the strong walls shook as he sang of Odin's deeds, and of him whom Odin's favour had raised on high, but who was now delivered over to the sword because he had drunk of the cup of madness. "Already," he said, "I see my favourite's sword stained with his blood. Now thou seest Odin himself. Arise if thou canst!" And Grimnir arose, the chains fell from his hands, the flames played harmlessly about his garments; he stood there in all his Ase's strength, his head surrounded by rays of heavenly light. Geiröd had at first half drawn his sword in anger; but now, when he tried to descend from his throne in haste to

ODIN BETWEEN TWO FIRES IN GEIRÖD'S PALACE.

attempt to propitiate the god, it slipped quite out of its sheath, he tripped over it and fell upon it, so that its blade drank in his heart's blood. After his death, Agnar ruled over the kingdom, and by the favour of Odin his reign was long and glorious.

ODIN, THE DISCOVERER OF THE RUNES, AND GOD OF POETRY AND WISDOM.

Odin's power and wisdom and knowledge are described in the Edda and in many of the lays of the skalds. He went to Mimir, the wise Jotun, who sat by the fountain of primeval wisdom, drank daily of the water and increased his knowledge thereby. The Jotun refused to allow the god to drink of his fountain, unless he first pledged him one of his eyes. Allfather did as he requested him, in order that he might create all things out of the depth of knowledge, and from that day forward Mimir drank daily of the crystal stream out of Allfather's pledge. Other accounts make out that the water was drawn out of Heimdal's Giallarhorn. Both accounts are given in the Northern poems. The myth from which they came shows us the meaning that lay at their foundation.

Mimir, a word related to the Latin *memor*, *memini*, signifies *memory;* that it was known to the Germans is indicated by the similar sounds of the names of the Mümling, a stream in the Odenwald, and of Lake Mumel in the Black Forest, where the fairies lived. Mimir drew the highest knowledge from the fountain, because the world was born of water; hence, primeval wisdom was to be found in that mysterious element. The eye of the god of heaven is the sun, which enlightens and penetrates all things; his other eye is the moon, whose reflection gazes out of the deep, and which at last, when setting, sinks into the ocean. It also appears like the crescent-shaped horn with which the Jotun drew the draught of wisdom.

According to other poems, Mimir was killed, but his head, which still remained near the fountain, prophesied future events. Before the Twilight of the Gods came to pass, Odin used to whisper mysterious things with him about the Destruction and Renewal of the world.

At one time when the god was standing with his golden helmet on, by the side of the holy fountain on the high hill, and learning the runic signs from Mimir's head, he discovered the Hugrunes (spirit-runes). As we have already shown, these runes were not exactly used as formulæ for writing connected sentences. They were only the accented letters used in Northern and Old-German poems; that is to say, they were letters of similar sound used for alliterative purposes. The following examples are some of those that remain to us from olden time: hearth and home; wind and weather; hand and heart. They were intended as a help to the memory when learning and singing the lays.

Odin gained power over all things by means of the runes, through which he was able to make all bend to his will, and to obtain authority over the forces of nature. He knew runic songs that were effectual in battle, in discord, and in time of anxiety. They blunted the weapons of an opponent, broke the chains of noble prisoners, stopped the deadly arrow in its flight, turned the arms of the enemy against themselves, and calmed the fury of angry heroes. When a bark was in danger on the stormy sea, the great god stilled the tempest and the angry waves by his song, and brought the ship safe to port. When he sang his magic strain, warriors hastened to his assistance and he returned unhurt out of the battle. At his command a man would arise from the dead even after he had been strangled. He knew a song that gave strength to the Ases, success to the elves, and even more wisdom to himself; another that gave him the love of woman so that her heart was his for ever more. But his highest, holiest song was never sung to woman of mortal birth, but was kept for the Queen of Heaven alone, when he was sitting peacefully by her side.

THE DRAUGHT OF INSPIRATION. ODIN'S VISIT TO GUNLÖD. JOURNEY TO WAFTHRUDNIR.

Kwasir, a man whom the Ases and Wanes had created amongst them, and whom they had inspired with their own spirit, was loved by gods and men for his wisdom and goodness. He travelled through all lands, teaching and benefiting the people. Wherever he went he tamed down the wild passions of all men, and taught them better and purer manners and customs.

The evil race of Dwarfs alone, they that burrowed in the earth in search of treasures, cared nought for the love, although they envied the wisdom of Kwasir. Fjalar and Galar, brothers of this people, invited him one day to a feast, and then murdered him treacherously with many wounds. They caught his blood in three vessels, the kettle Odrörir (inspiration), and the bowls Son (expiation) and Boden (offering). They mixed rum-honey with it, and made it into mead, which gave all who drank of it the gift of song and of eloquence that won every heart.

As the wicked deed of the Dwarfs had brought them such good luck, they invited the rich giant Gilling and his wife to visit them, and took the former out fishing with them. Then they upset the boat in the surf under great over-hanging rocks, so that Gilling was drowned, while they, being good swimmers, righted the boat again, and rowed to land.

When the giantess heard the sad fate of her husband, she wept and moaned, and refused to be comforted. The Dwarfs offered to take her to the rock on which the body had been washed. But as she was leaving the house, Galar threw a mill-stone from above down upon her head, so that she also was killed. Now Suttung, son of the murdered giant's brother, heard of the evil deed, and set out to avenge it. He seized the Dwarfs and made ready to bind them to a solitary rock out in the sea, that they might die there of hunger. They begged for mercy, promising to give him the wonderful mead concocted out of Kwasir's blood, in atonement for what they had done. The giant accepted the expiation offered him; he took the three vessels containing the liquor to a hollow mountain that belonged to him, and set his daughter Gunlöd to keep guard over the magic drink.

Odin, the God of Spirit, was told of all these things by his ravins Hugin and Munin. He determined to get possession of the Draught of Inspiration at any cost to himself, that it might no longer be kept uselessly hidden away by the giant in the interior of the earth, but might refresh gods and heroes, so that wisdom and poetry might delight the world. He therefore, in the guise of a simple traveller, started for Jotunheim. He came to a field where nine uncouth fellows were mowing hay. He offered to sharpen their scythes for them, and make them cut as well as the best swords. The men were pleased with his offer, so he pulled a whet-stone out of his pocket, and whetted and sharpened the scythes. When he at last returned them to the mowers, they found that they could work much quicker and better than before, and each wanted to have the whet-stone for himself. So the traveller threw it amongst them, and they struggled and fought for it with their scythes, until at length they all lay dead on the ground.

The traveller went on his way till he came to the master of the estate, the Jotun Baugi, a brother of Suttung, who received him hospitably. In the evening the giant complained that his farm-servants were all killed, and that his splendid crop of hay could not be harvested. Then Bölwerker (Evil-doer), as the traveller called himself, offered to do nine men's work if his host would get him a draught of Suttung's mead.

"If thou wilt serve me faithfully," answered the Jotun, "I will try to fulfil thy desire; but I will not hide from thee that my brother is very chary of giving a drop of it away."

Bölwerker was satisfied with this promise, and worked as hard as the nine farm-servants for the whole summer.

When winter came, Bangi, true to his promise, drove to his brother's dwelling with the traveller, and asked for a draught of the mead. But Suttung declared that the vagabond should not have a single drop.

"We must now try what cunning will do," said Bölwerker; "for I must and shall taste that mead, and I know many enchantments that will help me to what I want. Here is the mountain in which the mead is hidden, and here is my good auger, Rati, which can easily make its way through the hardest wall of rock. Take it and bore a hole with it, no matter how small."

The Jotun bored as hard as he could. He soon thought that he had made a hole right through the rock, but Bölwerker blew into it and the dust came out into the open air. The second time they tried, it blew into the mountain, and Bölwerker, changing himself into a worm, wriggled through the hole so quickly that treacherous Bangi, who stabbed at him with the auger, could not reach him.

When he had got into the cave, the Ase stood before the blooming maiden Gunlöd, in all his divine beauty and wrapped in his starry mantle. She nodded her acquiescence when he asked her for shelter and for three draughts of the inspiring mead.

Three days he spent in the crystal mansion, and drank three draughts of the mead, in which he emptied Odrörir, Son and Boden He was intoxicated with love, with mead, and with poetry. Then he took the form of an eagle, and flew with rhythmical motion to the divine heights, even as the skald raises himself to the dwellings of the immortals on the wings of the song that is born of love, of wine, and inspiration. But Suttung heard the flap of the wings and knew who had robbed him of his mead. His eagle-dress was

ODIN'S VISIT TO GUNLÖD.

at hand, he therefore threw it round his great shoulders, and flew so quickly after the Ase that he almost came up with him. The gods watched the wild chase with anxiety. They got cups ready to receive the delicious beverage. When Odin with difficulty reached the safe precincts of holy Asgard, he poured the mead into the goblets prepared for it. Since that time Allfather has given the gods the Draught of Inspiration, nor has he denied drops of Odrörir to mortal men when they felt themselves impelled to sing to the harp of the deeds of the gods and of earthly heroes.

Odin possessed knowledge of all past, present, and future events, since he had drunk of the fountain of Mimir and of Odrörir. He therefore determined to attempt a contest with Wafthrudnir, the wisest of the Jotuns, in which the conquered was to lose his head.

In vain Frigg strove, in her fear, to dissuade him from the perilous undertaking; he set out boldly on his way and entered the giant's hall as a poor traveller called Gangrader.

Stopping on the threshold of the banqueting hall, he said, "My name is Gangrader, I have come a long way; and now I ask thee to grant me hospitality and to let me strive with thee in wise talk."

Wafthrudnir answered him: "Why dost thou stand upon the threshold, instead of seating thyself in the room? Thou shalt never leave my hall unless thou hast the victory over me in wisdom. We must lay head against head on the chance; come forward then and try thy luck."

He now proceeded to question his guest about the horses that carried Day and Night across the sky, the river that divided Asgard from Jotunheim, and the field where the Last Battle was to be fought. When Gangrader had shown his knowledge of all these things, the giant offered him a seat by his side, and in his turn answered his guest's questions as to the origin of earth and heaven, the creation of the gods, how Niörder had come to them from the wise Wanes, what the Einheriar did in Odin's halls, what was the origin of the Norns, who was to rule over the heritage of the Ases after the world had been burnt up, and what was to be the end of the Father of the gods.

After Wafthrudnir had answered all of these questions, Gangrader asked: "I discovered much. I sought to find out the meaning of many things, and questioned many creatures. What did Odin whisper in the ear of his son before he ascended the funeral pile?"

Recognising the Father of the gods by this question, the conquered Jotun exclaimed: "Who can tell what thou didst whisper of old in the ear of thy son? I have called down my fate upon my own head, when I dared to enter on a strife of knowledge with Odin. Allfather, thou wilt ever be the wisest."

The poet does not tell us whether the visitor demanded the head of the conquered Jotun. Nor does he mention the word that Odin whispered to his son before he went down to the realms of Hel; but the context leads us to suppose that it was the word Resurrection, the word which pointed to the higher, holier life, to which Baldur, the god of goodness, should be born again, when a new and purer world should have arisen from the ashes of the old, sin-laden world.

ODIN, FATHER OF THE ASES.
ODIN'S DECENDANTS.

From later poems Odin appears not only as Ruler of the world, and Father of all Divine beings, who gradually as time went on became more and more subordinate to him, but also as progenitor of kings and heroic races, such as the kings of the Anglo-Saxons and Franks, as well as of the rulers of Denmark, Norway, and Sweden.

ODIN, FATHER OF THE GODS AND OF THE ASES.

According to the Edda, Odin had three sons, Wegdegg, the East Saxon; Beldegg (Baldur or Phol), the West Saxon (Westphalian); and Sigi, to whom Franconia was given; and three others, Skiöld, Säming, and Yngwi, who were made kings of Denmark, Norway, and Sweden. Other sagas show that Wals, Sigmund, and Sigurd, the hero of the Niflung Lay, were descended from Sigi, while Brand and Heingest or Hengist, Horsa and Swipdager were descended from Beldegg. The Anglo-Saxon genealogical tables make out that Voden (Wodan) and Frealaf (Freya) had seven sons, who were the founders of the Anglo-Saxon kingdom. Others, on the contrary, only show three sons here also, which makes them more in agreement with the northern genealogies.

According to the higher ideas regarding him, Odin was the father of gods and men; the latter were created by him, while the former were his direct or indirect descendants. His son by Jörd (the Earth) was strong Thor, father of Magni and Modi (Strength and Courage); by Frigg he had Baldur and Hödur; by Rinda, Wali, who afterwards became the avenger of Baldur; and by the nine mothers, the mysterious watchman Heimdal. Besides these, there were the poet-god Bragi; the divine messenger, Hermodur; the brave archer, Uller; and even the god of heaven, Tyr, who otherwise received the highest honours. Related to him were Forseti, son of Baldur, and Widar, who were to rule over the new world of holiness and innocence. Thus he was the Father of the Ases. On the other hand, Hönir, who gave to newly created man senses and life, and Loki, who gave him blood and blooming complexions, were Odin's brothers or comrades in primeval times. Great Niörder, his bright son Freyer and his daughter Freya belonged to another divine race, that of the Wanes; they were first brought into Asgard as hostages, but were received into the ranks of the Ases.

FRIGG AND HER MAIDENS.

After the birth of Thor, whose mother was Jörd (the Earth), daughter of the giantess Fiörgyn, Odin left the dark Earth-goddess and married bright Frigg, a younger daughter of Fiörgyn; henceforth she shared his throne Hlidskialf, his divine wisdom and his power, becoming the joy and delight of his heart, and the mother of the Ases. She ruled with him over the fate of mortals and granted her votaries good fortune and victory, often bringing about her ends by woman's cunning. Just as in Hellas a feast was held each year in commemoration of the marriage of Zeus and Hera, so did the old Teutons in like manner hold festivity to celebrate the union of Odin and Freya.

Freya's palace was called Fensaler, that is, the hall of the sea. It probably got this name from the dwellers on the coast, who looked upon Frigg as the ruler of the sea and protector of ships. A soothing twilight always reigned, and it was adorned with pearls and gold and silver. And the goddess would bring all lovers, and husbands and wives who had been separated by an early death, to this peaceful palace, where they were reunited for ever. This belief of the old Teutons shows us that they regarded love in its truest and highest aspect, and built their hopes on being reunited after death to the objects of their affections. What we learn from the Latin annals of Armin and Thusnelda, of the high position of women as seers of future events, proves to us that noble women were always treated even by rude, fighting men, with respect and reverence; while the romance of love is clearly shown in the Northern myth of Brynhild, who threw herself upon the burning pyre in order that she might be reunited to her beloved Sigurd.

In her gorgeous palace Frigg sits spinning, on her golden distaff,

FRIGG AND HER MAIDENS.

the silken threads, which she afterwards bestows on the most worthy housewives. The goddess' spinning-wheel was visible to man every night, for it was that shining, starry zone which we in our ignorance now point out as the Belt of Orion, but which to our ancestors was the Heaven-queen's spinning-wheel. The goddess had three friends and attendants always beside her, and with these she used to hold council on human affairs, in the hall of the moon.

Fulla or Volla was the first of Frigg's attendant-goddesses, and chief of the maidens; according to Teutonic belief she was also the sister of the Queen of Heaven. She wore a golden circlet round her head, and beneath it her long hair floated over her shoulders. Her office was to take charge of the Queen's jewels, and to clothe her royal mistress. She listened to the prayers of sorrowful mortals, repeated them to Frigg, and advised her how best to give help.

Hlin, the second of Frigg's maidens, was the protector of all who were in danger and of those who called upon her for help in hour of need.

The messenger of the Queen of Heaven was Gna, who rode, swift as the wind, on a horse with golden trappings, over land and sea, and through the clouds that floated in the air, to bring her mistress news of the fate of mortal men.

Once as Gna was hovering over Hunaland, she saw King Rerir, a descendant of Sigi and of the race of Odin, sitting on the side of a hill. She heard him praying for a child, that his family might not be blotted out of memory; for both he and his wife were advanced in years, and they had got no child to carry on their noble race. She told the goddess of the prayer of the king, who had often presented fine fruit as a sacrifice to the heavenly powers. Frigg smilingly gave her an apple which would ensure the fulfilment of the king's desire. Gna quickly remounted her horse Hoof-flinger, and hastened over land and sea, and over the country of the wise Wanes, who gazed up at the bold rider in astonishment, and asked:

> "What flies up there, so quickly driving past?"
> Her answer from the clouds, as rushing by:
> "I fly not, nor do drive, but hurry fast
> Hoof-flinger swift through cloud and mist and sky."

King Rerir was still seated on the hillside under the shade of a fir-tree, when the divine messenger came down to earth at the skirt of the wood close to where he sat. She took the form of a hooded-crow, and flew up into the fir-tree. She heard the prince mourning over the sad fate that had befallen him, that his family would die out with him, and then she let the apple fall into his lap. At first he gazed at the fruit in amazement, but soon he understood the meaning of the divine gift, took it home with him and gave it to his spouse to eat.

Meanwhile Gna guided her noble horse rapidly along the star-lit road to Asgard, and told her mistress joyously of the success of her mission. In due time the Queen of Hunaland had a son, the great Wolsing, from whom the whole family took its name. He was the father of brave Sigmund, the favourite of Odin, and he in his turn of Sigurd, the fame of whose glory was spread over every Northern and Teutonic land.

When the Queen of Heaven heard of the success that had accompanied her divine gift, she herself decided to be the bearer of the news to the assembled gods and heroes, and determined to appear in her most glorious array. Fulla spread out all the Queen's jewels until they shone like stars, yet Frigg was not satisfied. Then Fulla pointed to Odin's statue of pure gold, that stood in the hall of the temple. She thought a worthy ornament might be made for the goddess out of that gold, if the skilful artificers who had made such a marvellous likeness of the Father of the gods could only be won over. The artists were bribed with rich presents and they at last cut away some of the gold from a place that was covered by the folds of the floating mantle, so that the theft could not easily be discovered. They then made the Queen a necklace of incomparable beauty. When Frigg entered the assembly and seated herself on the throne beside Odin, she at once made known to all present how she had saved a noble family from extinction. Every one gazed at her beauty in amazement, and the Father of the gods felt his heart filled anew with love for his queen.

A short time afterwards Odin went to the hall of the temple in which his statue was placed. His penetrating eye at once discovered the theft that no one else had noticed, and his wrath was immediately kindled. He sent for the goldsmiths, and as they confessed nothing, he ordered them to be executed. Then he commanded that the statue should be placed above the high gate of the temple, and prepared magic runes that should give it sense and speech, and thus enable it to accuse the perpetrator of the deed. The Goddess-queen was greatly alarmed at all these preparations. She feared the anger of her lord, and still more the shame of her deed being proclaimed in the presence of the ruling Ases.

Now there happened to be in the Queen's household a serving demon of low rank, but bold and daring, who had already ventured to show his admiration for his mistress. Fulla went to him and assured him that the Queen was touched by his devotion, upon which the demon declared himself willing to run any risks for her sake. He made the temple watchmen fall into a deep sleep, tore down the statue from above the door, and dashed it in pieces, so that it could no longer speak or complain.

Odin saw what he was doing and guessed the reason. He raised Gungnir, the spear of death, ready to fling at all who had been concerned in the evil deed. But his love for Frigg triumphed over all else ; he determined on another punishment.

He withdrew from gods and men ; he disappeared into distant regions, and with him went every blessing from heaven and earth. A false Odin took his place, who let loose the storms of winter and the Ice-giants over field and meadow. Every green leaf withered, thick clouds hid the golden sun and the light of the moon and stars ; the earth, lakes and rivers were frozen by the raging cold which threatened to destroy all forms of life. Every creature longed for the return of the god of blessing, and at length he came back. Thunder and lightning made known his approach. The usurper fled before the true Odin ; and shrubs and herbs of all kinds sprouted anew over the face of the earth, which was now made young again by the warmth of spring.

In the foregoing tale, we have endeavoured as much as possible to make a conneeted narrative out of the confused, and now and then contradictory, myths regarding Frigg and her handmaids. We will only add that the myth which completes it, dates from a time when the gods had paled in the eyes of the people, and had become less exalted in character than of old. There are many versions of it differing from one another, and it serves here to show the difference between Summer-Odin and Winter-Odin.

OTHER GODDESSES RELATED TO FRIGG.

Let us now again turn our attention to the great goddess Frigg. The Northern skalds first raised her to the throne and distinguished her from Freya or Frea, the goddess of the Wanes. She was originally identical with her, as her name and character show. For Frigg comes from *frigen*, a Low-German word connected with *freien* in High-German, and meaning to woo, to marry, thus pointing to the character of the goddess. The old Germanic races, therefore, knew Frea alone as Queen of Heaven, and she and her husband Wodan together ruled over the world. The name Frigga or Frick was also used for her, for in Hesse, and especially in Darmstadt, people used to say fifty years ago of any fat old woman : " Sie ist so dick wie die alte Frick." (She is as thick [fat] as Old Frick.) The word *frigen* is also related to *sich freuer* (rejoice) ; thus Frigg was the goddess of joy (*Freude*). She took the place of the Earth-goddess Nerthus (mistakenly Hertha), who, Tacitus informs us, was worshipped in a sacred grove on an island in the sea. Nerthus was probably the wife of the god of heaven, in whom we recognise Zio or Tyr. He was the hidden god who according to the detailed account of Tacitus, was so reverently worshipped in a sacred grove by the Semnones, the noblest of the Swabian tribes, that the people never set foot on the ground that was consecrated to him without having their hands first bound. The Earth-goddess may also have been the wife and sister of Niörder, and separated from him when he was received amongst the Ases. In this case she belonged to the earlier race of gods, the Wanes, and her husband must have then been called Nerthus, a name afterwards changed into Niörder.

In Mecklenburg the same goddess appears under the name of Mistress Gaude or Gode, which is the feminine form of Wodan or Godan. The country people believed that she brought good luck with her wherever she went.

One story informs us that she once got a carpenter to mend a wheel of her carriage, which had broken when she was on a journey. She gave him all the chips of wood as a reward for his trouble. The man was angry at getting so paltry a remuneration, and only pocketed a few of the chips; but next morning he saw with astonishment that they had turned to pure gold.

According to another tale, Dame Gode was a great huntress,

who together with her twenty-four daughters devoted herself to the noble pursuit of the chase day and night, on week-days and on Sundays. She was therefore made to hunt to all eternity, and her pack of hounds consisted of maidens who were turned into dogs by enchantment; she was thus forced to take part in the Wild Hunt.

In France the goddess was called Bensocia (good neighbour, *bona socia*), and in the Netherlands, Pharaildis, *i.e.*, Frau Hilde or Vrouelden, whence the Milky Way was named Vrouelden-straat.

Hilde (*Held*, hero) signifies war, and she was a Walkyrie, who with her sisters exercised her office in the midst of the battle. Later poems make her out to be daughter of King Högni, who was carried off, while gathering magic herbs on the seashore, by bold Hedin when he was on a Wiking-raid. Her father pursued the Wiking with his war-ships, and came up with him on an island. In vain Hilde strove to prevent bloodshed. Högni had already drawn his terrible sword, Dainsleif, the wounds made by which never healed. Once more Hedin offered the king expiation and much red gold in atonement for what he had done.

His father-in-law shouted in scorn: "My sword Dainsleif, which was forged by the Dwarfs, never returns to its sheath until it has drunk a share of human blood!"

The battle began and raged all day without being decided one way or the other.

In the evening both parties returned to their ships to strengthen themselves for the combat on the morrow.

But Hilde went to the field of battle, and by means of runes and magic signs awakened all the dead warriors and made whole their broken swords and shields.

As soon as day broke, the fight was renewed, and lasted until the darkness of night obliged the combatants to stop.

HILDE, ONE OF THE WALKYRIES.

The dead were stretched out on the battle-field as stiff as figures of stone; but before morning dawned the witch-maiden had awakened them to new battle, and so it went on unceasingly until the gods passed away.

Hilde was also known and worshipped in Germany, as is shown by the legend about the foundation of the town of Hildesheim.

One year, as soon as snow had fallen on the spot dedicated to her, King Ludwig ordered the cathedral to be built there. The Virgin Mary afterwards took her place, and several churches were built in honour of Maria am Schnee (Marie au neige) both in Germany and in France.

Nehalennia, the protectress of ships and trade, was worshipped by the Keltic and Teutonic races in a sacred grove on the island of Walcheren; she had also altars and holy places dedicated to her at Nivelles. The worship of Isa or Eisen, who was identical with Nehalennia, was even older and more wide-spread throughout Germany. St. Gertrude took her place in Christian times, and her name (Geer, *i.e.*, spear, and Trude, daughter of Thor) betrays its heathen origin.

HOLDA. OSTARA.

Once upon a time, in a lonely valley of the Tyrol, where snow-capped glaciers ever shone, there lived a cow-herd with his wife and children. He used to drive his small herd of cattle out to graze in the pastures, and now and again would shoot a chamois, for he was a skilled bowman. His cross-bow also served to protect his cattle from the beasts of prey, and the numerous bear-skins and wolf-skins that covered the floor of his cottage bore witness to his success as a hunter.

One day, when he was watching his cattle and goats on a fragrant upland pasture, he suddenly perceived a splendid chamois, whose horns shone like the sun. He immediately seized his bow and crept forward on hands and knees until he was within shot. But the deer sprang from rock to rock higher up the mountain, seeming every now and then to wait for him, as though it mocked his pursuit. He continued the chase eagerly until he reached the glacier which had sunk below the snow-fields.

The chamois now vanished behind some huge boulders, but at the same time he discovered a high arched doorway in the glacier, and in the background beyond he saw a light shining.

He went through the dark entrance boldly, and found himself in a large hall, the walls and ceiling of which were composed of dazzling crystal, ornamented with fiery garnets. He could see flowery meadows and shady groves through the crystal walls; but a tall woman was standing in the centre of the hall, her graceful limbs draped in glancing, silvery garments, caught in at the waist by a golden girdle, and resting on her blond curls was a coronet of carbuncles. The flowers in her hand were blue as the eyes with which she gently regarded the cow-herd. Beautiful maidens, their heads crowned with Alpine roses, surrounded their mistress, and seemed about to begin a dance. But the herdsman had no eye for any except the goddess, and sank humbly on his knees.

Then she said in a voice that went straight to the heart of the hearer :—

"Choose what thou thinkest the most costly of all my treasures, silver, gold, or precious stones, or one of my maidens."

"Give me, kind goddess," he answered; "give me only the bunch of flowers in thy hand; I desire no other good thing upon the earth."

She bent her head graciously as she gave him the flowers, and said :—

"Thou hast chosen wisely. Take them and live as long as these flowers bloom. And here," pointing to a corn measure, "is seed with which to sow thy land that it may bear thee many blue flowers such as these."

He would have embraced her knees, but a peal of thunder shook the hall and the mountain, and the vision was gone.

When the cow-herd awoke from his vision, he saw nothing but the rocks and the glacier, and the wild torrent that flowed out of it; the entrance to the palace of the goddess had vanished. The nosegay was still in his hand and beside him was the wooden measure full of seed. These tokens convinced him that what had happened was not a mere dream.

He took up his presents and his cross-bow, and descended the mountain thoughtfully to see what had become of his cattle. They were nowhere to be seen, look for them where he might, and when he went home he found nothing but want and misery. Bears and wolves had devoured his herd, and only the swift-footed goats had escaped from the beasts of prey.

A whole year had elapsed since he had left home, and yet he had thought that he had only spent a few hours chamois-hunting in the mountains. When he showed his wife the bunch of flowers, and told her that he intended to sow the seed that had been given him, she scolded him, and mocked him for his folly; but he would not be turned aside from his determination, and bore all his wife's hard words most patiently.

He ploughed up a field and sowed the seed, but there was still a great deal over; he sowed a second and a third field, and yet much seed remained. The little green sprouts soon showed in the fields, grew longer and longer, till at length the blue flowers unfolded themselves in great numbers, and even the cow-herd's wife rejoiced at the sight, so lovely were they to look upon.

The man watched over his crop day and night, and he often saw the goddess of the mountain wandering through his fields in the moonlight with her maidens, blessing them with uplifted hands.

When the flowers were all withered and the seed was ripe, she came again, and showed how the flax was to be prepared, after which she went into the cottage and taught the cow-herd's wife how to spin and weave the flax and bleach the linen, so that it became as white as newly fallen snow.

The cow-herd rapidly grew rich, and became a benefactor to his country, for he introduced the cultivation of flax throughout the land, which gave employment and wages to thousands of country-people. He saw children, grandchildren, and great-grandchildren around him, but the bunch of flowers the goddess had given him was still as fresh as ever, even when he was more than a hundred years old and very tired of life.

' One morning while he was looking at his beloved flowers, they all bent down their heads, withered and dying. Then he knew that it was time to say farewell to earthly life. Leaning on his staff, he toiled painfully up the mountains. It was already evening when he reached the glacier.

The snow-fields above were shining gloriously as though in honour of the last walk of the good old man. He once more saw the vaulted doorway and the glimmering light beyond. And then he passed with good courage through the dark entrance into the bright morning which greets the weary pilgrim, when, after his earthly journey is over, he reaches Hulda's halls. The door now closed behind him, and he was seen no more on earth.

This and other traditions of the same kind are told in the Tyrol of the old Germanic goddess Hulda or Holda. Her name shows that she was a goddess of grace and mercy, and she must have been worshipped both in Germany and in Sweden, but still no traces are to be found of her at the present day in the Teutoburg Forest, where so many of the places and names point back to the old Germanic religion, nor yet do the Northern skalds give an account of her. However, German fairy legends and tales call to

HOLDA, THE KIND PROTECTRESS.

us the great goddess whose character and deeds live on in the memory of the people, and the Northern Huldra, who drew men to her by means of her wondrous song, is exactly identical with her. Her name has been derived from the old Northern Hulda, *i.e.*, Darkness; and it has been thought that she was the impersonation of the dark side of the goddess of Earth and Death; but the derivation which we gave before, from Huld, grace, mercy, seems more suitable.

A Northern fairy-tale makes Hulla or Hulda, queen of the Kobolds. She was a daughter of the queen of the Hulde-men, who killed first her faithless husband and then herself. She enticed wise King Odin by means of a stag, to her mansion, which was hidden in the depths of a wood. She gave him of her best, and then begged him to act as umpire in a legal dispute that had arisen between her and the other Kobolds and Thurses, about the murder of her husband. He consented to do so, and his decision made her queen of all the Kobolds and Thurses in Norseland. This tale is quite modern in its form, but it certainly is based on ancient beliefs.

A poem dating from the middle ages places Holda in the Mountain of Venus, a place that is generally supposed to be the Hörselberg in Thuringia. She was then called Mistress Venus, and held a splendid court with her women. Noble knights, amongst whom was Ritter Tannhäuser, were drawn by her into the mountain, where they lived such a gay, merry life of pleasure that they could hardly ever again free themselves from her spell and make their escape, even though thoughts of honour and duty might now and then return to them.

It was finally said of Holda, that those who were crippled in any way were restored to full strength and power by bathing in her Quickborn (fountain of life), and that old men found their vanished youth there once more. This tradition connects her with the

Northern Iduna, who had charge of the apple that preserved the immortality and vigour of the Ases. But she also resembled Ostara, who was worshipped by the Saxons, Franks and other tribes.

Ostara, the goddess of Spring, of the resurrection of nature after the long death of winter, was highly honoured by all the old Teutons, nor could Christian zeal prevent her name being immortalised in the word Easter, the period of spring, at which time the Saxons in England worshipped her. The memory of these old times has long since passed away, although the "hare" still lays its "Easter-eggs." The custom is very old of giving each other coloured eggs as a present at the time when day and night became equal in length and when the frozen earth awakens to new life after the cold of winter is gone, for an egg was typical of the beginning of life. Christianity put another meaning on the old custom, by connecting it with the feast of the Resurrection of the Saviour, who, like the hidden life in the egg, slept in the grave for three days before he wakened to new life.

There are no legends about the goddess of spring. One monument alone, and that a newly discovered one, remains of the old worship, the Extern-stones, which are to be found in the Teutoburg Forest at the northern end of the wooded hills. It is stated in the chronicle of a neighbouring village, dating from last century, that the ignorant peasantry were guilty of many misdemeanours there when doing honour to the heathen goddess Ostara. Had the clergyman only told us whether there were processions, dances, feasts, scattering of flowers, or any other kind of sacrifice, a clear light might have been shed over the manner in which the goddess was worshipped. Still, this fact proves that not only the name, but also the worship of Ostara was kept in the memories of the people for hundreds, perhaps thousands, of years, and shows how deeply rooted it was. The rocks may perhaps have been called Eastern or Eostern-stones, and may have been dedicated to Ostara. There, as elsewhere, the priests and priestesses of the goddess probably assembled in heathen times, scattered Mayflowers, lighted bonfires, slaughtered the creatures sacrificed to her, and went in procession on the first night of May, which was dedicated to her. Very much the same as this used to be done at Gambach, in Upper Hesse, where, as late as thirty years ago even, the young people went to the Easter-stones on the top of a hill, every Easter, and danced and held sports. Edicts were published in the eighth century forbidding these practices; but in vain, the people would not give up their old faith and customs. Afterwards the priestesses were declared to be witches, the bonfires, which cast their light to great distances, were said to be of infernal origin, and the festival of May was looked upon as the witches' sabbath. Nevertheless, young men and maidens still continue, near the Meissner-Gebirg in Hesse, to carry bunches of Mayflowers and throw them into one of the caves that are to be found there. For Ostara, who gives new life to nature, is the divine protectress of youth and the giver of married happiness.

BERCHTA OR BERTA.

The dusk of evening has fallen over Berlin. A great yet silent crowd is rapidly moving through the chief street towards the royal palace, and every now and then a low whisper is heard, in which can be distinguished the words: "The King is very ill." In the palace itself yet greater silence reigns. The King's guardsmen stand motionless, the servants' steps are inaudible on the carpets of the corridors and the rooms. Now the tower clock strikes midnight; all at once a door opens, and through it glides a ghostly woman, tall of stature, queenly of bearing.

She is dressed in a trailing white garment, a white veil covers her head, below which her long flaxen hair hangs, twisted with strings of pearls; her face is deathly pale as that of a corpse. In her right hand she carries a bunch of keys, in her left a nosegay of Mayflowers. She walks solemnly down the long corridor. The tall guardsmen present arms, pages and lackeys give way before her, the guards who have just relieved their comrades open their ranks; the figure passes through them, and goes through a folding door into the royal ante-room.

"It is the White Lady; the King is about to die," whispers the officer of the watch, brushing a tear from his eye.

"The White Lady has appeared," is whispered through the crowd, and all know what that portends.

At noon the King's death was known to all. "Yes," said Master Schneckenburger, "he has been gathered to his fathers. Mistress Berchta has once more announced what was going to happen, for she can foretell everything, both bad and good. She was seen before the misfortunes of 1806, and again before the battle of Belle-Alliance. She has a key with which to open the door of life and happiness. He to whom she gives a cowslip will succeed in whatever he undertakes."

Schneckenburger was right. It was Bertha, or Berchta, who made known the King's approaching death, but she was also the prophetess of other important events. Berchta (from *percht*, shining) is almost identical with Holda, except that the latter never appears as the White Lady. Many Germanic tribes worshipped the Earth-goddess under the name of Berchta, and there are numbers of legends about her both in North and South Germany.

One evening in the year was dedicated to her, and was called Perchten-evening (30th December or 6th January), when she was supposed, as a diligent spinner, to oversee the labours of the spinning-room, or, magic staff in hand, to ride at the head of the Raging

Host, in the midst of a terrific storm. She generally lived in hollow mountains, where she, as in Thuringia, watched over and tended the "Heimchen," or souls of babes as yet unborn, and of those who died an early death. She busied herself there by ploughing up the ground under the earth, whilst the babes watered the fields. Whenever men, careless of the good she did them, disturbed her in her mountain dwelling, she left the country with her train, and after her departure the fields lost all their former fruitfulness.

Once when Berchta and her babes were passing over a meadow across the middle of which ran a fence that divided it in two, the last little child could not climb over it; its water-jar was too heavy.

A woman, who a short time before had lost her little baby, was close by, and recognised her dead darling, for whom she had wept night and day. She hastened to the child, clasped it in her arms, and would not let it go.

Then the little one said: "How warm and comfortable I feel in my mother's arms; but weep no more for me, mother, my jar is full and is growing too heavy for me. Look, mother, dost thou not see how all thy tears run into it, and how I've spilt some on my little shirt? Mistress Berchta, who loves me and kisses me, has told me that thou shouldst also come to her in time, and then we shall be together again in the beautiful garden under the hill."

Then the mother wept once more a flood of tears, and let the child go.

After that she never shed another tear, but found comfort in the thought that she would one day be with her child again.

Berchta appears in many legends as an enchantress, or as an enchanted maiden, who provided a rich treasure for him who was lucky enough to set her free from the magic spell that bound her. Still more frequently, however, she took up her abode in princely castles as the "Ahnfrau," or Ancestress of the family to whom the castle belonged. In these stories the Goddess of Nature is hardly recognisable.

It is told that the widowed Countess Kunigunde of Orlamünd fell in love with Count Albrecht the beautiful, of Hohenzollern. He told her that four eyes stood in the way of a marriage between them, and she, thinking that he referred to her children, had them secretly murdered. But, as the tale informs us, he had meant his parents, who disapproved of the marriage. He felt nothing but abhorrence of the murderess when he found out what she had done, and she, repenting of her sin, made a pilgrimage to Rome, did severe penance, and afterwards founded the nunnery of the Heavenly Crown, where she died an abbess. Her grave, as well as those of her children and of the Burggraf Albrecht, are still shown there. From that time she appeared at the Plassenburg, near Baireuth, as the "Ahnfrau," who made known any evil that was going to happen; later on she went to Berlin with the Count's family, and is still to be seen there as the tale at the beginning of this chapter shows.

Another account makes the apparition out to be the Countess Beatrix of Cleve, who was married to the Swan-Knight so often mentioned among the old heroes of the middle ages. The House of Cleve was nearly related to that of Hohenzollern, and in the mysterious Swan-Knight we recognise the god of Light, who comes out of the darkness of night and returns to it again.

A more simple version refers to a Bohemian Countess, Bertha of Rosenberg. She was unhappily married to Johann of Lichtenberg, after whose death she became the benefactress of her subjects, built the Castle Neuhaus, and never laid aside the white garments of widowhood as long as she lived. In this dress she appeared, and even now appears, to the kindred families of Rosenberg, Neuhaus and Berlin, on which occasion she prophesies either good or evil fortune.

The Germanic races carried the worship of this Earth-goddess with them to Gaul and Italy, in the former of which countries a proverbial expression refers to the underground kingdom of the goddess, by reminding people "*du temps que Berthe filait.*" It was that time of innocence and peace, of which almost every nation has its tradition, for which it longs, and to which it can only return after death.

Historical personages have also been supposed to enact the part formerly given to the Earth-mother.

A tradition of the 12th century informs us that Pepin, father of Charlemagne, wished to marry Bertrada, a Hungarian princess, who was a very good and diligent spinner. His wooing was successful, and the princess and her ladies set out on their journey to Pepin's court. The bride's marvellous beauty was only marred by her having a very large foot.

Now the chief lady-in-waiting was a wicked woman, and jealous of Bertrada; so she gave the princess to some villains she had bribed, in order that she might be murdered in the forest, and then she put her own ugly daughter in her mistress's place. Although Pepin was disgusted with his deformed bride, he was obliged to marry her according to compact; but soon afterwards, on finding out the deception that had been practised upon him, he put her from him.

Late one evening when out hunting, he came to a mill on the river Maine. There he saw a girl spinning busily. He recognised her as the true Bertrada by her large foot, found out how her intended murderers had taken compassion on her, and how she had finally reached the mill. He then discovered his rank to her, and entreated her to fulfil her engagement to him. The fruit of this marriage was Charlemagne.

In this tale we recognise the old myth under a modern form.

We see how Mother Earth, the protectress of souls and ancestress of man, especially of those of royal or heroic race, is thrust aside by the cunning, wintry Berchta, but is joined again by her heavenly husband, and becomes the mother of the god of Spring. Even the large foot reminds us of the goddess, who was originally supposed to show herself in the form of a swan. This is the reason why in French churches there are representations of queens with a swan's or goose's foot (*reine pédauque*).

Other French stories show Berchta in the form of Holda: how she sheds tears for her lost spouse, so bitter that the very stones are penetrated by them. Both goddesses are identical with the Northern Freya, who wept golden tears for her husband.

There is an old ballad that is still sung in the neighbourhood of Mayence, which tells of the bright, blessed kingdom of the goddess. We can give only the matter of it here, as the verses themselves have not remained in our memory.

A huntsman once stood sadly at the water's edge, and thought on his lost love. He had had a young and lovely wife, who, when he came wearied home from the chase, would welcome him with the warm kiss of love. She bare him a sweet babe, and made him perfectly happy. But ere long both were taken from his side by grim, envious death, and now he was alone. Gladly would he have died with them, but that was not to be. Three months had flown by, but his wife and child were still always in his thoughts.

One night his way led him beside a flowing stream; he stopped still on the bank, gazed long into the water's depths, and asked:

"Is the broken heart to be made whole in a watery grave alone?"

Thereupon sweet silvery notes fell upon his ear; and as he glanced upwards, he saw before him a beauteous, queenly woman, sitting opposite him on the other side of the stream; she was spinning golden flax, and singing a wondrous song:

"Youth, enter thou my shining hall,
    Where joy and peace e'er rest ;
When the weary heart at length finds all
    Its loved ones, 'gain 'tis blest !

The coward calls my hall the grave,
    My kiss he fears 'twere death ;
But the leap is boldly made by the brave—
    His the gain by the loss of life's breath !

Youth, leave thou, then, the lonesome, des'late shore,
And boldly gain the joy enduring evermore."

The huntsman listens ; do the thrilling tones come from the beauteous woman on the opposite bank, or is it from the watery deep that they proceed ?

Wildly he leaps into the flood, and a fair, white arm is extended, encircling him and drawing him down beneath the water's surface, away from all earthly cares, away from all earthly distress and pain. And his loved ones greet him, his youthful wife and his babe. "See, father ! how green the trees grow here, and how the coloured flowers sparkle with silver ! And no one cries here, no one has any troubles ! "

This tale is based upon the old heathen belief as to the life in a future state ; it shows us that the conviction of our forefathers has always been, that for the virtuous death was merely a transition to a new life, to a life purer, more complete, than that on earth.

THOR, THUNAR (THUNDER).

Arwaker (Early-waker) and Alswider (All-swift), the horses of the sun, were wearily drawing the fiery chariot to its rest. The sea and the ice-clad mountains were glowing in the last rays of the setting sun. The clouds that were rising in the west received them in their lap. Then flashes of lightning darted forth from the clouds, thunder began to roll in the distance, and the waves dashed in wild fury upon the rock-bound coast of the fiord.

"Hang up the snow-shoes, lad, and take off thy fur cap; Öku-thor (Thor of the chariot) is driving over to waken old Mother Jörd. Put the jar of mead on the stone table, wife, that he may find something to drink; and you, you lazy fellows, why are you sitting idly over the fire, instead of rubbing up the ploughshares until they shine again? This is going to be a fruitful year, for Hlorridi (heat-bringer) has come early. Come, Thialf, pull off my fur boots."

Thus spoke the yeoman to whom Balshoff belonged, as he sat on the stone bench by the fire. But then he stopped short, and stared open-mouthed; Thialf let the fur boots fall from his hand; the mistress of the house dropped the jug of mead, and the farm-servants the plough. Wingthor drove over from the west in all his fury; he struck the house with his hammer Miölnir, and the flash broke through the ridge of the roof beside the pillar that supported it, and penetrated a hundred miles below the clay floor. A sulphureous vapour filled the room; but the yeoman, shaking off his stupefaction, rose from his stone bench, and when he saw that no more damage was done, he said:

"Wingthor has been gracious to us, and now he has gone on to fight against the Frost and Mountain Giants. Do ye not hear the blows of his hammer, the howls of the monsters in their caverns, and the crashing of their stone heads as though they were nothing but oatmeal dumplings? But to us he has given rain, which even now is falling heavily, rain that will soon melt away the snow and prepare the soil to receive the seed we shall sow later on. The tiny sprouts will grow rapidly, and grass and herbs and the green leek will reward us for our industry. Preserve the golden ears of corn for us, O Thor, until the harvest time."

In such manner people used, in the olden time, to call on the strong god of thunder, Thunar,—in the North, Thor. He was held in great reverence, and was perhaps even regarded as an equal of the God of Heaven. Traces of this are still recognisable, for wherever he was spoken of in connection with the other gods, he was given the place of honour in the middle. The Saxons had to renounce Wodan, Donar, and Saxnot. In the temple of Upsala, Thor is placed between Odin and Freyer. In " Skirnir's Journey," a poem of the Edda, it is said : " Odin is adverse to thee, the Prince of the Ases (Thor) is adverse to thee, Freyer curses thee." He retained this high position in Norway, where he fought against the Frost and Mountain Giants, who sent the destructive east wind over the country. And not less honour was paid him in Saxony and Franconia. The oak was sacred to him, and his festivals were solemnized under the shade of oak trees. When thunder-clouds passed over the earth, Thor was said to be driving his chariot drawn by two fierce male goats, called Tooth-cracker and Tooth-gnasher.

Odin—not he who sat on Hlidskialf overlooking the nine worlds, but the omnipotent God of Heaven—married Jörd, Mother Earth and the offspring of this marriage was strong Thor, who began even in the cradle to show his Ase-like strength by lifting ten loads of bear-skins.

Gentle old Mother Jörd, who was known by several other names in different parts of Germany, could not manage her strong son, so two other beings, Wingnir (the winged), and Hlora (heat) became his foster-parents. These were personifications of the winged lightning. From them were derived the god's names of Wingthor and Hlorridi.

Thor married Sif (kin), for he, the protector of households, was himself obliged to have a well-ordered household. The beautiful goddess had golden hair, probably because of the golden corn of which her husband was guardian, and her son was the swift archer,

Uller, who hunted in snow-shoes every winter, and ruled over Asgard and Midgard in the cold season, while the summer Odin was away. By the giantess, Jarnsaxa (Ironstone) Thor had two sons, Magni (Strength) and Modi (Courage), and by his real wife a daughter, Thrud (Strong), the names of whom all remind us of his own characteristics.

Thor was handsome, large and well-proportioned, and strong. A red beard covered the lower part of his face, his hair was long and curly, his clothes were well-fitting and his arms were bare, showing his strongly-developed muscles. In his right hand he carried the crashing-hammer, Miölnir, whose blows caused the destructive lightning flash and the growling thunder.

THOR AND LOKI'S JOURNEY IN WOMEN'S CLOTHES.

THOR'S DEEDS AND JOURNEYS.

THE MAKING OF MIÖLNIR.

A gentle breeze was blowing over the rich land of Thrudheim, and the doors of Bilskirnir were standing open that the castle might be filled with the aromatic perfume of the summer flowers. Thor slept quietly in the great hall, until morning dawned and chased away the shades of night. The god then rose from his couch, but his first glance fell on his wife Sif, who looked very sad. All her golden hair had vanished in the night, and she was standing before him with a bald head, like the earth when the golden corn has been harvested. He guessed who the author of the mischief was, and rushed angrily over the hills and through the groves of Asgard until he came to spiteful Loki, whom he seized by the throat and held till his eyes almost started from his head. He would not let him go until he promised to obtain another head of hair, the same as the old one, from the dwarfs. As soon as the mischief-maker was free he hastened to Elfheim, and after paying a heavy price, brought away with him not only the hair but also Gungnir, the spear that never failed in its blow; and the ship Skidbladnir, which could sail whatever wind was blowing, and which was so cunningly made, that it could be folded up and put in the pocket when it was no longer wanted. He gave Thor the hair for his wife, and it was no sooner put upon her head than it took root and began to grow apace. To Odin he gave the spear, and to Freyer the ship, that he might go to sea with the merchants' galleys and save shipwrecked persons.

Delighted with the praise his gifts received on all sides, Loki asserted that his smiths, the sons of Iwaldur, were the best workers in metal that had ever lived. Now it happened that the Dwarf Brock was present when he said this, and Brock's brother, Sindri, was generally regarded as the best smith. So he scornfully replied that no one could beat his brother, and that he would wager his head for Sindri's fame. Brock informed his brother of the dreadful bet, but was told to be of good courage; he was given the bellows and desired to keep on blowing the fire without stopping, so that there might be no interruption in the magic work, a circumstance which would at once bring all their efforts to naught. Sindri then put a pig-skin in the fire, and went away to draw the magic circle, and command the assistance of the hidden powers in his labours. Brock, meanwhile, worked hard at the bellows, in spite of the attacks of a fly which continually stung him on the hand till the blood flowed. When Sindri returned there was life in the fire, and he drew out of it the enormous wild boar Gullinbursti, with golden bristles, the radiance of which made the dark smithy as light as day.

The second work of art had now to be made. Sindri laid some red gold in the furnace, and Brock blew the bellows in spite of the cruel stings of the fly, until at last the ring Draupnir was formed, from which eight other rings exactly similar dropped every ninth night.

Lastly, the smith threw a bar of iron into the furnace, and desired his brother to blow steadily. Brock did as he was told, and bore the agony caused by the fly, which he knew cunning Loki had sent. But when all at once it stung him on the eyelid, and the blood ran down into his eye, he dashed his hand at it to crush it. Then the flames rose in the air and suddenly sunk again and were extinguished. Sindri rushed into the hall in terror, but his face brightened when he had looked into the furnace.

"All is well," he said; "it is finished—only the handle is somewhat short."

Then he drew a great battle-hammer out of the furnace, and gave it to his brother, as well as the two other works of art, adding:

"Go now; thou hast won the bet, and thine enemy's head also."

Brock entered the assembly of the Ases, who were sitting in council. He gave Odin the ring Draupnir, and to bright Freyer he gave the boar Gullinbursti, which he said would carry him swift as the wind through mists and clouds, and over mountains and valleys. When Thor received the hammer, and swung it in his right hand, then he, the prince of the Ases, grew tall as a giant; dark clouds piled themselves around his waist; lightning flashed from the clouds, and rolling peals of thunder shook the heights of Asgard and Midgard, terrifying both Ases and mortal men. Odin alone, to whom fear was impossible, sat unmoved upon his throne, and said;

"Miölnir is the greatest of treasures, for in the hand of my son it will protect Asgard from every assault of the giants."

So Brock won the wager and Loki's head as well, and he refused to accept anything else in exchange. But the son of Laufey had already taken refuge in flight, so Thor hastened after him, and soon brought him back.

"The head is thine, but not the neck," cried the mischief-maker, as the dwarf raised his sword.

"Then I will sew up thy great mouth," answered Brock, trying to make holes through his opponent's lips; but all in vain, the knife made no impression. So he got his brother's awl, and that did not fail. He sewed up the mouth, and Loki stood in the midst of the laughing Ases unable to speak; yet he soon found means to unfasten the string.

The hair of the earth-goddess, Sif, is the flowers and corn that grow upon the earth. These are cut down in the harvest, and the winter-demon robs the goddess of her hair, and leaves her head quite bald. But the Dwarfs who live under the earth provide her with a fresh supply of hair, and with the help of the Thunder-god punish the evil-doer.

Alwismal, the Song of Alwis.—Alwis, the King of the Dwarfs, who had travelled throughout the nine worlds and had learnt all the languages and wisdom of the dwellers therein, once went to Asgard. He met with a friendly reception there, for all the Ases knew about his palace which shone with gold and precious stones, and of his widely extended power over the underground people. He saw beautiful Thrud, Asathor's strong daughter, fell in love with her, and asked for her hand in marriage. The Ases approved of the proposal of the King of the underground treasures, and were of opinion that Thor would be pleased with the arrangement. So the marriage day was fixed. But Thor came home before the wedding-day, and was very wroth when he was told the news.

"Who art thou, thou pasty-faced fellow?" he asked of the would-be bridegroom; "Hast thou been with the dead? Hast thou arisen from the grave to snatch the living back with thee to thy dismal kingdom?"

Alwis now asked him who he was that pretended to have power over his bride and to be able to prevent the marriage which was already arranged; but when he found that it was Wingthor, Thrud's father, he told him of his possessions and of his wisdom, and entreated him to consent.

Thor, in order to prove him, asked what certain words were in the different languages of men, Ases, Wanes, Jotuns, Elves, and in Helheim.

The Dwarf answered everything right; but lo! day began at that moment to break, and Alwis was touched by a ray of sunlight, whereupon he stiffened into stone, and remained on the heights of Asgard, a monument of Thor's victory.

THOR'S JOURNEY TO UTGARD.

The Hrimthurses sent out cold winds from the interior of Jotunheim over the fields of Midgard, so that the tender green shoots were blighted and the harvest spoilt. Thor, therefore, ordered his chariot to be got ready, and hastened away to force the giants to keep within bounds. Loki joined him with flattering speeches, and the Thunderer thought that it might be as well to take him with him, as he knew his way about the wilderness so well.

Thor's goats went so quickly that the travellers reached the bare rocks of the giants' country by the evening.

They saw a lonely farmhouse, and the owner offered them hospitality, but could only give them a poor supper. Thor, therefore, slew his goats and boiled them in a pot. He then invited his host and all his people to join him at supper, but commanded them to throw all the bones on the skins which he had spread out on the floor, and to beware how they broke any.

Cunning Loki whispered to the farmer's son, Thialfi, that he ought to break one of the thigh bones, as the marrow in it was good to eat. Thialfi followed the evil counsel, and found that the marrow was indeed most excellent.

Next morning Thor waved his hammer over the skins and bones, and immediately the goats jumped up, but one of them was lame in the hind leg. The god was very angry, his eyes flashed, his right hand closed round the handle of his hammer, and a thunderclap shook the house to its foundations. The farmer, who had been flung upon his face, begged for mercy, and his wife and children joined him in his entreaties; he offered his son Thialfi and his daughter Röskwa in atonement for the broken thigh-bone.

Then the angry god grew calm, and accepted the expiation offered him; he left his goats and chariot behind and walked on with his companion and the sturdy children of the farmer towards Jotunheim.

They crossed high mountains, and went through deep valleys until they came to a broad sound. When they had crossed the sound, their way led them over a stony country and through a dark wood that seemed as if it would never end. The ground was covered with a grey mist, out of which an iceberg, resembling a corpse-like ghost, here and there reared its head. All was dim and uncertain, as though surrounded by enchantment.

The travellers pursued their journey all day long, Thialfi, the quickest runner in the country, always keeping in front with Thor's travelling bag.

In the evening they reached a strange, roomy inn, in which there was neither inhabitant nor food to be found; yet they lay down to rest, as they felt very hungry.

At midnight a violent earthquake shook the house, but they

succeeded in finding a place within the building that seemed to be more secure than the rest; there Thor's companions took refuge, whilst he, hammer in hand, kept watch by the entrance. Loud sounds of roaring and snorting disturbed the sleep of the travellers. The Prince of the Ases awaited the morning.

When it grew light, he perceived a man of mighty stature, whose snoring had been the cause of all the noise they had heard. He felt very much inclined to bless the snorer's sleep with a goodly blow of his hammer, but at that very moment the giant awoke.

In reply to his question, "Who art thou?" the giant answered that his name was Skrymir, and added that he knew perfectly well that his questioner was Asathor. As he said this, he began to look about for his glove. And how great was the astonishment of the Ase, when he discovered that he and his companions had spent the night in the giant's glove, and that when they had been startled out of their first resting-place, they had taken refuge in the thumb.

Skrymir gave himself no further trouble about the surprise of the strangers, but laid out his breakfast and devoured it, whilst the travellers took some provisions for themselves out of Thor's bag. The giant then tied up all his belongings in a bundle, threw it over his broad back, and walked on before the others through the wood at such a pace that they could hardly follow him. In the evening they took up their quarters for the night under an oak tree, the top of which reached the clouds.

The Jotun gave the travellers the remains of the food in his bundle, because, he said, sleep was more necessary for him than food. The strong Thunderer vainly strove to unfasten the cord tied round the bundle. Enraged by this failure, he pulled his girdle of strength tighter round his waist, and seizing Miölnir with both hands, dealt a terrible blow on the head of the snoring giant, who merely rubbed the place with his hand, and asked whether a leaf had fallen on his head.

At midnight the wood again re-echoed with his snores. Thór now hit the monster again as hard as he could on the crown. The hammer made a deep hole, but Skrymir thought that it was only an acorn that had fallen upon him, and soon began to snore again.

Towards morning the angry Ase dealt a third dreadful blow at the giant; the earth trembled, rocks fell with a horrible crash; the hammer penetrated the giant's skull, so that the end was hidden. Nevertheless, Skrymir rose quietly and said :—

"So, thou art awake already, Asathor. Look, some birds, when building their nests, have let a little bit of stick fall on my temple; it is bruised. We must part here; my way lies to the north, and yours to Utgard in the east. You will soon see Utgard-Loki's castle before you. There you will find bigger men than I. Beware lest any of you open your mouths too wide in boastful talk; for if you do, you will get into difficulties."

Skrymir went straight on through the wood, while the others turned in the direction he had pointed out to them.

About noon they came in sight of the giant's castle, which was large and shining as an iceberg. They slipped in between the bars of the postern gate, and entered the royal hall.

There sat Utgard-Loki, Prince of the Thurses, on his throne, and ranged around him on benches were his warriors and courtiers. He stared at the travellers in surprise.

"I know ye well, little people," he cried, in a voice that resembled the rumbling of a falling rock. "I know thee, Asathor, and guess that thou canst do more than thy appearance would justify one in supposing. Now tell me what each of you can do, for no one is allowed to sit down here without showing himself to be good for something."

First of all Loki vaunted his powers in eating.

SKRYMIR ATTACKED BY THOR WHEN ASLEEP.

"A good thing to be able to do on a journey," said the King; "for then one can eat enough at one meal to last for eight days. Logi, my cook, shall try with thee which is the better trencherman. We shall see which of you can eat the most."

A large trough was filled with meat, and the two heroes stood one at each end of it, and tried which could devour the fastest. They met in the middle; Loki had eaten one half of the meat, and Logi the other; but as the latter had at the same time disposed of the bones and the trough as well, he walked away from the table proud of his victory.

Thialfi announced that he was swift of foot, and challenged the courtiers to race with him in the lists. A young fellow named Hugin accepted the challenge. He turned back at the goal just as the farmer's son reached it.

"Well run for a stranger, by my beard," growled the Prince of the Thurses; "but now make better speed." However, Thialfi was farther behind at the second turn, and at the third he had full half the course to run when Hugin turned at the goal.

It was now time for Thor to show what he could do. He first said that he could drink a long draught. The Thurse commanded that the horn should be brought that some could empty at one draught, many at two, and the weakest at three. The Ase looked at the horn. It was long, but it was narrow, and he thought he could easily dispose of the contents. Nevertheless, the first draught hardly uncovered the rim, the second very little more, and the third a few inches at most. Much ashamed, he gave back the horn; he could drink no more.

He then spoke of his strength. Utgard-Loki told him to pick up the grey cat which was lying purring at his feet. The hammer-thrower imagined that he could fling the cat up to the ceiling; but his first attempt to lift it only made it arch its back, at the second it arched its back a little more, at the third he raised one paw from the ground; farther than that he could not move it. He heard with rage the scornful laughter with which his fruitless efforts were greeted from the benches. Lightning flashed from his eyes; he challenged the courtiers to wrestle with him in the lists.

"That will go ill with thee," said the King, stroking his beard: "try first what thou canst do here against Elli, my old nurse; she has conquered stronger men than a shrimp like thee before now."

The old woman was ready by this time, and seized strong Thor, who exerted all his strength to try and overthrow her. But she stood as immovable as a rock, and used her own strength so well, that he sank upon one knee.

"Enough," cried the Jotun. "Sit down, strangers, and enjoy my hospitality."

On the following morning the king accompanied them as far as the wood.

"Here," he said, "are the borders of my domain, which you should never have crossed had I known more about you. Let me now tell you how I have tricked you. Three times, Asathor, didst thou strike at my head; but I always shoved a mountain between me and thee. Look, dost thou see the marks made by thy hammer, three deep abysses, the last of which reaches down to the Home of the Black-Elves? The cook Logi, who measured his strength against Loki, and who devoured even the bones and the trough, was wild-fire. Hugin, was Thought, whom neither Thialfi nor any other runner could expect to overtake. The drinking horn was connected with the ocean. Thou didst drink so much that every shore was left uncovered, and the people said: 'It is ebb tide.' Thine eyes were blinded when thou didst lift the grey cat, for then thou didst swing the Midgard-snake as high as heaven, and she had nearly wriggled herself free and done irreparable injury. Elli, the nurse, who looked so weak, was old age, which none can withstand when his time has come. Go now, for this is my realm, where I have dominion over the Hrimthurses and their rocky fastnesses. Where I rule, there is no space for men to cultivate the land, yet Asathor might split the mountains and the eternal ice with his thunder."

Thor had already raised his hammer to punish the Jotun for his magic spells, but he had vanished. A bare, stone-strewed wilderness surrounded him and his companions. Columns of mist hovered here and there, out of which Jotuns were peering, now with a smile of scorn and again looking down grimly, now sinking and again rising in the air, so that the travellers did not know what was real and what enchanted. They then set out on their return to Thrudheim.

The natural myth which gave rise to this poem of the Younger Edda is very suitable for our collection. Not even the mighty Ase could make it possible for man to cultivate the soil amongst the great mountains, where rock is piled upon rock, and all are covered with ice and snow. Thialfi is the diligence which must animate the farmer, and his sister Röskwa is the quickness and activity which must attend him.

Duel with Hrungnir.—Thor passed some happy days in his halls of Bilskirnir. His fair wife Sif, who kept the house in good order, was beautiful as the May moon; her artistically-made golden hair grew daily longer, and fell over her neck and shoulders in ringlets. The god had great pleasure in his son Magni, who, although only three years old, was as tall and strong as a man. The Jotuns in the neighbourhood were all quiet, for they did not care to harm the husbandmen's crops. Still, the farmers who lived far away in valleys amid the inhospitable mountains, often called upon the helpful Ase to defend them against the monsters, who sent storms, floods, avalanches, and falling rocks, to disturb them in their peaceful labours. Thor then hastened with Miölnir to punish the peace-breakers in the east.

Allfather Odin was away on his travels, now ruling the battle of mortal men, now searching after wisdom, and now wooing the favour of women with loving words. Upon one of these journeys he arrived at the castle of the Mountain-giant, Hrungnir, where he was hospitably received. Whilst they were talking together, the Jotun remarked that Sleipnir was a good horse, but that his own horse, Gullfaxi (golden mane), was better, and that it could leap farther with its four feet than the former with its eight.

"Well," cried Odin, "I will wager my head upon my horse. Catch me if thou canst."

He jumped upon Sleipnir and galloped away, the giant pursuing him with a giant's rage.

Swift as the storm-wind, the Father of the gods galloped on far ahead. Hrungnir was not aware, in his haste, that his golden-maned horse was thundering over the bridge Bifröst until he stopped at the gates of Walhalla. Then the King of the Ases came out to meet him, and in return for his hospitality led him into the hall. To Hrungnir was given the enormous goblet, full of foaming beer, from which Thor was accustomed to drink. In his ill-humour, he emptied it in a few draughts, and asked in his intoxication for more and more.

"Ha!" he exclaimed, "none of you know me yet. I will take Walhalla upon my back and carry it off to Jotunheim. I will throw Asgard into the abyss of Nifelhel, and strangle you all, except Freya and Sif, whom I will take home with me. I will empty all your beer barrels to the sediment. Bring me what you have. Freya shall be my cup-bearer."

The trembling goddess poured him out a bumper, but the other Ases called aloud for Thor.

The god appeared in the hall with the speed of the lightning that flashes down from the sky.

"Who has permitted the Thurse to sit down in holy Asgard?"

he demanded in a voice of thunder. "Why does Freya give him the drinking-horn? His head shall be broken in punishment for this."

And as he said these words, his eyes sparkled and his hand closed round the shaft of his hammer.

Then Hrungnir immediately at once became sober. He stammered out that Odin had invited him to the feast, and that it would be dishonourable of Thor to attack an unarmed man. Yet he would be ready to fight with him at Griottunagard (rollingstone, or also rock-wall) in the borders of Jotunheim.

The Ase could not withdraw from this challenge, and the Jotun made all the haste he could to reach home with a whole skin.

Everywhere and in all countries the coming duel was talked about. The Jotuns knew that their best fighting man was going to venture on a dangerous undertaking. They consulted together how they might ensure him the victory.

They made a clay man nine miles high and three miles across the chest, Möckerkalfi (Mist-wader) by name, who was to help their hero in the fight, but who had only a trembling mare's heart in his breast. The Jotun himself had a triangular heart of stone, and his skull was also of stone, and his shield and his club too.

Hrungnir and his clay squire awaited Thor at Griottunagard on the appointed day. The Ase did not waste time. He drove up in the midst of rolling thunder and flashing lightning, surrounded by clouds. His quick-footed servant, Thialfi, ran on before him, and called out to the Jotun that he was mistaken in holding his shield before him, for the god would come up out of the ground to attack him.

Then Hrungnir flung his shield under his feet and seized his club in both hand, to be in readiness to throw it, or to hit out with it. He now perceived the Ase swinging Miölnir, so he threw his club at him with fearful strength. The weapons crashed together in the middle of the lists; but the force of the hammer was so great that it splintered the club and broke the stone-head of the giant in pieces, felling him almost dead to the ground. Meanwhile a splinter from the club had penetrated Thor's forehead, so that he also fell, and as it happened, right under the leg of the falling giant. Sturdy Thialfi had in the meantime despatched the clay giant with a spade, and had broken him up into the clay from which he had been made. He now tried to help his master, but could not lift the giant's leg. Other Ases tried also, until at length the strong boy Magni came up. And he pushed aside the heavy weight as though it were a mere trifle, saying:

"What a pity it is, Father, that I did not come sooner; I could have broken that fellow's stone head with my fist."

"Thou wilt be a strong man," said Thor; "and thou shalt have the good horse Gullfaxi as a reward for helping me."

He then strove to pull the stone splinter out of his brow, but could neither move it nor could he even loosen it, so he was forced to drive home to Thrudheim with an aching head.

Loving Sif and anxious Thrud vainly endeavoured to alleviate the pain Thor was enduring. The prophetess Groa (green-making) now came to the house. She could move rocks with her magic spells, and also stop the course of wild floods. She offered to cure Thor. Then she drew her circles and sang her wondrous songs. The stone began already to shake and grow looser, and the wounded Ase hoped for a speedy cure. In order to give Groa pleasure, he told her, while she murmured her spells, that he had waded across the ice-stream Eliwagar, carrying her husband, Örwandil, on his back, and had broken off one of Örwandil's frost-bitten toes, which he had flung up into the sky, where it was now shining like a star.

"And now," he said, "he is on his way home to thee."

Scarcely were the words out of his mouth, when Groa sprang up joyfully, forgetting all about her magic spells. And so the splinter remained in Thor's forehead.

According to the poet Uhland, this is a poetical description of the splitting of the rocks by the crashing hammer of the god. Thialfi, the diligent husbandman, conquered the clay giant, the uncultivated ground, while Thor made agriculture possible among the rocks. He was hurt by the falling stones when doing this. Groa (the green-making), the sprouting power in plants, was married to Örwandil (living seed), whom Thor carried on his shoulders through the wintry ice-streams Eliwagar. Mannhardt looks upon Örwandil as lightning sparks. We refrain from noticing further the different interpretations put upon the story. The skald found the natural myth, touched the strings of his harp and sang his song with all his heart, careless whether he gave the old myth in all its particulars or not.

Journey to Hymir.—In this myth the terrors of the polar regions are described. It was in that northern realm that the Frost-giant Hymir (the dusk-maker) ruled, and in his house lived the golden, white-browed goddess of light, who had been stolen from her home, and also the nine-hundred headed grandmother, the mountains of ice and snow.

Hymir was guardian of the great brewing vat, whose depth might be counted by miles; by this was probably meant the Arctic Ocean, through which the summer god, Thor, opened a passage for seafaring men. Thor conquered the terrors of the Arctic climate before which even the bold Wikings drew back appalled, while in our days, brave North Pole voyagers face them undauntedly.

Thus Uhland explains the myth, and we feel inclined to agree with him; nevertheless, this journey to Hymir is said by other commentators to mean a descent into the Under-world. Perhaps both explanations are admissible, for all nature is dead in winter, buried under a pall of snow, and the ideas of winter and death are frequently interchangeable. Strong Thor, therefore, descended into the Under-world, conquered its terrors, as he did those of the Hrimthurses, and returned home victorious, in like manner as Herakles did in the Greek myth, which ascribes to him a heroic deed of the same kind as this.

THOR'S JOURNEY TO THRYMHEIM TO GET BACK HIS HAMMER.

Night with her starry diadem had spread her mantle over Asgard. Every creature was asleep; the Ases in their golden chambers, and the Einheriar stretched out on the benches of Walhalla after a goodly feast on the flesh of Sährimnir, and many a draught of delicious mead. They dreamt happy dreams of brave deeds and of the joys of victory.

Wingthor alone tossed restlessly about on his cushions of down. He heard in his dreams the murmur of wicked runes, and saw a gigantic hand seize hold of Miölnir. At length he was awakened by hollow peals of thunder. He snatched at the hammer which always lay by his bedside, but could not find it. Angrily he sprang to his feet and felt about for it; but it was gone; the faint light of morning showed that the place where he had laid it was empty. He shook his head wrathfully and his eyes flashed fire. His beard grew redder than ever, and the house trembled at his shout:

"Miölnir is gone; it has been stolen by enchantment."

Loki heard his cry, and said to him:

"I will get thee back thy hammer, whoever has stolen it, if Freya will lend me her Falcon-dress."

So they went to Folkwang and entered the presence of Freya. They addressed her in courteous words, and asked her to lend them her feather-garment, that they might spy out who had stolen Miölnir.

And the gentle goddess answered : "You may have it. I would lend it to you willingly, even if it were made of silver or gold."

She then took the dress out of a chest and gave it to the Ases. And now Loki flew with rhythmic strokes of his wings, high above the precincts of Asgard and the swift river Ifing, until he reached the barren mountains of Jotunheim.

Thrym, a prince of the Thurses, was sitting there on a hill. He was decorating his dogs, that ran quickly as the wind, with golden ribbons, and making the manes of his fiery horses shine.

"What news dost thou bring from Asgard, that thou comest alone to Thrymheim ? " he called out to the new-comer : " how goes it with the Ases and how with the Elves ? "

"Badly with both Ases and Elves," answered Loki, "for Miölnir is lost. Speak, hast thou hidden it anywhere ? "

Then the Thurse laughed, and said : "I have hidden it eight miles deep in a cleft of the earth ; and no one shall have it unless he brings me Freya as a bride to my halls."

Enraged at his message, Loki flew back over the Ifing river to Asgard, where Thor awaited him. He gave the message of the wicked Thurse.

Again Thor and Loki went to visit the goddess in her shining hall at Folkwang.

"Up and dress thyself, Freya," said Thor ; "put on thy snowy bridal garments, and I will take thee to Thrym, prince of the Thurses."

Then the goddess' anger was kindled at this address, and she started from her throne, making the palace shake to its foundations.

"You may call me mad," she cried, "if ever I follow thee in bridal array to Thrymheim, to the Prince of the Thurses, monster that he is."

Having thus spoken, she dismissed the Ases from her presence without a word of farewell.

The Ases now all assembled on their seats of justice near the fountain of Urd, that they might consult together as to the best means of rescuing the hammer from the power of the Giants.

The first to speak was Heimdal, the god who resembled a Want in wisdom; he said :—

"Let Thor himself put on the bridal garments, let a bunch of keys jingle at his waist, let precious stones sparkle upon his neck, let his knees be covered by the petticoats of a woman, and a veil be put before his face.

The Prince of the Ases did not approve of the advice of wise Heimdal. He would, he said, be always called a woman in future, if he ever put on female apparel. But when Loki replied that if he did not get back the hammer the giants would soon come to live in Asgard, he consented to do as the Ases entreated.

Soon afterwards he sat in his chariot dressed as a bride, and Loki, son of Laufey, in the guise of a serving maid, seated himself by his side.

The goats set off; they rushed in wild leaps through Asgard and Midgard; the earth smoked, and rocks and mountains split with loud reports wherever they went.

Thrym was sitting comfortably at the threshold of his hall. He watched his golden-horned cows coming home, he saw his large herds of black bullocks, his stores of gold and precious stones in their iron caskets.

"I have a great store of riches," he said; "the only thing wanting now is that Freya should be my wife. And to-morrow she will enter my halls; so strew the benches my men, and have plenty of food and mead in readiness, for it beseems a spacious hall like mine that the wedding should be a merry one."

Early next morning the visitors arrived, and soon afterwards his bride was sitting beside Thrym, well-veiled, as modesty and custom demanded.

The tables were laden with costly food and wine, which were a pleasure to look at as well as to eat and drink. No one could rival the bride, however. She ate a fat ox in no time, then eight huge salmon, and all the sweet cakes that were made for the women, and in addition she drank two barrels of mead. The Thurse was astonished at her hunger.

"Well," he exclaimed, "I never before saw a bride with such an appetite, nor did I ever see a girl drink mead in such a degree!"

But the serving maid assured him that her mistress had tasted neither bite nor sup for a week, so excited had she been at the thought of her wedding.

The Jotun wished to kiss his bride on hearing this, and raised her veil for the purpose; but at the sight of Freya's flaming eyes, which seemed as though they flashed fire at him, he shrank back to the end of the room.

But the wise maid calmed down his apprehensions. "My lady," she said, "has not slept for a week, and that is the reason her eyes are so fiery."

The gaunt sister of the Thurse now approached the bride to ask for a wedding present.

"Give me," she entreated, "golden rings and a pair of buckles, and thou shalt enjoy my love."

Unmoved by this appeal, the bride sat silent in her wedding array. Then the Prince, intoxicated with love and mead, commanded that the hammer should be brought from its hiding-place, that the marriage might be solemnized in the usual way.

"And then," he added, "place it in the lap of the bride."

It seemed at that moment as though the bride were stifling a laugh beneath her veil, and indeed a ferocious laugh was heard when the Prince's command had been obeyed.

Now the bride rose, and threw off her veil; it was Asathor, terrible to look upon; he raised his bare arm and held Miölnir aloft in his mighty right hand. The walls of the room tottered and cracked, a peal of thunder shook the house and a flash of lightning darted through the hall. Thrym lay stretched on the floor with a broken head; his guests and his servants fell under the blows of the hammer; not even his gaunt sister escaped. The flames made their way out through the roof; and house and hall fell with a loud crash. A smoking heap of ruins alone remained to show the place where the powerful Thrym had ruled.

The spring sun rose; it shone down upon the devastated dwelling, the broken rocks, fallen stones, torn and uprooted soil, and upon the victorious god who had conquered the power of the enemy.

The storm-clouds of anger were gone from Thor's brow. He stood upon the height and gazed at his work of destruction with a gentle and kindly look upon his face. Then he called his children of men to come and instil new life into the destruction, so that farms and dwelling houses, agriculture and commerce, civic order, law and morality should arise and flourish there. And so into this conquered land came farmers and builders, with hatchet, spade, and plough; herdsmen with their cattle and sheep, and mighty hunters to keep down the numbers of bears and wolves. And Thor was in the midst of them, setting up stones to mark the boundaries, consecrating the tilled land with his hammer; then the grateful people erected an altar to him, made a great feast in his honour, and promised him the first-fruits of their labour. After that Thor got into his chariot, followed by Loki, and together they returned to Asgard rejoicing in what they had done.

We have pointed here to the natural myth which lies at the foundation of this poem. The myth is one of the most beautiful in the Elder Edda. The poet has made free use of the materials that were at his disposal, so that the most minute details of the primitive myth can never be discovered; yet the following can be made out with certainty.

The beneficent Thunder god, who ruled over summer, was deprived of his hammer in the winter; Thrym (Thunder) hid it eight miles deep in the ground, *i.e.*, for eight months. He desired to have possession of Freya, the fair goddess of spring, in order that he might deprive man of the bright weather she brought with her. But Thor regained his hammer, and slew the Frost-giant and his followers, and his gaunt sister too, who according to Uhland was the famine that haunts rude mountain districts. Thus the god opened a new field to human industry.

JOURNEY TO GEIRÖD'S-GARD.

Loki once took Frigg's falcon-dress; he wrapped himself in it and hovered over many an abyss and broad stream until he had flown right above the barren rocks and ice of Jotunheim. He saw a chimney in the distance, out of which fire and smoke were issuing. Quickly he flew there, and perceived that the chimney belonged to a rambling grange.

This was Geiröd's-Gard, where Prince Geiröd, the Hrimthurse, dwelt with his people. The Ase was curious to know what was going on in the large hall, and fluttered down close to the window. But the Thurse caught sight of the falcon, and sent a servant out to catch it. Loki amused himself by making the man climb the high railing above which he fluttered, taking care to keep, as he thought, just out of reach; but suddenly he was caught by the leg and given to the giant.

"This is a strange-looking bird," said Geiröd, staring into the falcon's eyes as though he thought he could thus discover its character. "Tell me," he asked, addressing it, "whence thou comest, and what thou really art?"

But the bird remained silent and motionless.

So the Prince determined to tame him through hunger, and locking him up in a chest left him there for three months without food.

When he was taken out at the end of that time, Loki told who he was and begged to be set free.

At this the Thurse laughed so loud that he shook the hall and the whole grange.

"At length," he exclaimed, "I have got what I have long desired, a hostage of the Ases. I will not let thee go until thou hast sworn a holy oath to bring me Thor, the Giant-killer, without his hammer and girdle of strength, that I may fight him hand to hand. I expect that I shall conquer him as easily as I would a boy, and then I shall send him down to Hel's dark realm."

Loki promised with a holy oath to do as the giant bade, and flew quickly away.

When the cunning Ase had recovered from his fatigue, he remembered his oath. He told strong Thor that Geiröd had received him most hospitably, and that he had expressed a great wish to see the unconquerable protector of Asgard face to face, but without the terrible signs of his power, of which he was much afraid. Loki went on to say that there were strange things to be seen at the giant's house which were not to be seen elsewhere. Thor listened to the tempter, and at once set out on his journey, accompanied by Loki.

On his way to Geiröds-gard he met the giantess Grid, by whom Odin had once had a son named Widar, the silent. She told him what the true character of Geiröd was, and lent him her girdle of strength, and her staff and iron glove as a defence against the giant.

The day after this, he and Loki reached the broad river Wimur, which stretched out before them like a sea, and was so wide that the other shore was invisible. When Thor began to wade across, steadying himself by means of his staff, the water rose, and the waves beat wildly against his shoulders.

"Do not rise, Wimur," he cried, "for I must wade over to the giant's house."

Then he saw Geiröd's daughter, Gialp, standing in the cleft of a rock and making the water rise. He forced her to flee by throwing a great stone at her, and afterwards got safely over to the other bank, which he managed to climb, swinging himself up by means of a service tree. Loki also got safely over, for he clung to Thor's girdle the whole way.

When the travellers saw the chimney with the fire issuing from it, and the castle high as a mountain just in front of them, they knew that they had got to the end of their journey.

They went into the entrance hall. Thor seated himself wearily upon the only chair that was to be seen. But he soon discovered that it was rising higher and higher, so that he was in danger of being crushed against the ceiling. He pressed the end of his staff against the beams that ran across the top of the hall, and with all his Ase-strength tried to force the chair down again. A terrible crack and a cry of pain told him that he had hurt some living creature in his struggles. Gialp and Greip, Geiröd's daughters, had raised the chair on which he was sitting, and they now lay under it with broken backs, victims of their own cunning.

A monster serving-man now challenged Thor to a fencing bout in the great hall. On entering it the Ase saw with amazement that fires were burning all round the walls, the flames and smoke of which rose through the chimney he had seen before.

Instead of giving him courteous greeting, the Jotun king flung an iron wedge at him, which he had taken red hot out of the furnace with a pair of tongs. But Thor caught it in his iron glove and threw it back with such impetus that it broke through the brazen breastplate and body of the Jotun, and then crashed through the wall, burying itself deep in the earth on the other side of it. Thor looked down on the cowering giant who had at once turned into stone. He set him up as a monument of his victory, and there the petrified monster remained for centuries, reminding succeeding generations of men of the great deeds done by Asathor.

This is said to be another of the natural myths which tell how the beneficent god of summer conquered the destructive tempest with his own weapons; the two daughters are supposed to be personifications of the mountain torrents which caused rivers to overflow.

According to some, however, this legend, like the last one, describes a descent of the god into the Underworld, and there is also a similar one related by Saxo Grammaticus, of which Thorkill is the hero.

But we are of opinion that it is far more likely to have been in the volcanic island of Iceland that Thor was victorious over the demon. The island was known to the skalds, from the descriptions of bold sailors, long before its colonization by the Northmen. Tales of volcanic eruptions and hot springs must have excited the imagination of the poets extremely. Thus perhaps arose the myth of Thor's journey to Geiröds-gard, in which the god conquers the demon of subterranean fire. This view is supported by the shape of a rock near Haukadal, where, within a circle of 900 feet, are geysers and strocks. The rock is said to resemble a gigantic man cowering down, his body broken in the middle.

THE HARBARD LAY.

In this poem Odin acts the part of a ferryman, under the name of Harbard, refuses to row Thor, the god of agriculture, over the river, and sends him on his way with opprobrious words.

The reason was, that Odin was the god of the spirit and the warlike courage which animated the nobles and their retainers The proud warriors and skalds despised the peaceful peasantry who remained quietly at home, lived upon herrings and oatmeal porridge, and hated the devastation caused by war; while they, on the contrary, were continually fighting for wealth and glory, and hoped to rise to Odin's halls after death upon the field of battle.

This contempt for the tiller of the soil is clearly shown in the Lay, which makes the protector of agriculture play a very pitiful part. The myth had its rise in later times, when the old faith in the gods and deep reverence for them had already begun to decay.

The bold Wikings did not hesitate to say that they trusted more in their own good swords than in the help of Odin and Asathor. The Lay was perhaps composed at that time, but still, it rested on an older one, in which the myth of agriculture, of the apparent death of Fiörgyn or Jörd, mother of Thor, through the devastation caused by war, and of the renewed life of the Earth-goddess, were more clearly described.

IRMIN.

As we have before remarked, the Prince of the Ases was worshipped as one of the holy ones by the Teutonic race; it is probable that he was also adored under the name of Irmin, and that the different Irmin-columns were dedicated to him. But Irmin means universal, and it was to the universal, omnipotent god that the Irmin-columns were erected. It was he who helped the Teutons to victory in their battles against the Romans; for this reason the celebrated Irmin-column, which was destroyed nearly 800 years later by Charlemagne, was set up in his honour at Osning (in the Teutoburg Forest). It also reminds us of the hero Armin, who was held in great reverence, and whose name and character were in process of time confounded with those of the god.

Irmin was also supposed to be identical with the mythical hero Iring, who, when the Franks and the Saxons were fighting against the Thuringians, traitorously slew his lord, Irminfried, and then killed the false-hearted ruler of the Franks. After this he cut his way through the ranks of the enemy, sword in hand, and did many other heroic deeds. If this hero was the same as Irmin, he was very different from Thor, whose nature in all the myths regarding him was always true-hearted, and never cunning. But the legend also makes out the traitor to have been different from the god, for, after their victory, the Saxons erected a pillar to Irmin, and not to the Thuringian Iring.

Irmin was the common god of many tribes, and some philologists derive the name "German" from him. He was the guardian deity of the Thuringians, Katti, and Cherusci, and showered down his blessings upon them as he drove over the firmament of heaven in the Irmin-wain (Great Bear or Charles' Wain). The Milky-way, Iring or Irmin-road, the way of souls, was also sacred to him, and thus he was the ruler of souls, and identical with Aryama, the national god of all the Aryan races in the oldest times. The Kelts worshipped the same god under the names of Erimon and Erin, whence Ireland and the Irish are called after him. The chariot in which he drove through the heavens showed

CHAINING OF THE FENRIS WOLF.

his relationship to Thor according to the oldest ideas; but still Odin, the Leader of souls, had much in common with him. Tyr, the ancient god of heaven, the sword-god, was, however, yet more nearly kin to him, because he was depicted in warlike array, and because the monuments of victory, the Irmin-columns, were called after him. Several places have also derived their names from him.

TYR OR ZIO.

Who is there, who, after a hard day's work, has not rejoiced to see the approach of quiet Mother Night, when, wrapped in her starry mantle, she brings back peace to the world which has been robbed of it by restless Day?

This feeling of peace has often been destroyed by a sound that has something mysterious and strange about it. It is only the long-drawn howl of a dog, a sound that is heard most frequently when the moon is shining brightly; but it has something gruesome in it, and this accounts for the popular belief that it betokens the death of the person who hears it.

A circumstance of this kind happened once upon a time within the holy precincts of Asgard.

Mani (the moon) was following Mother Night merrily in his chariot, when suddenly he started and his happy face became clouded, for out of a great abyss there arose a howling noise which quickly swelled to a dreadful roar, so that the whole earth trembled as after a peal of thunder.

The Ases were awakened by it, and the Einheriar snatched at their weapons, for they thought that Ragnarök had come. Amongst them stood Tyr, tall and slender as a pine, and unmoved by the terrors that they had expected.

"Fenris," he said, "has been wakened by the moon, and wants something to eat; I will go and feed him."

Then he set out in the night, laden with living and dead animals with which to appease the monster's rapacity. Once more the terrible roar was heard, then it seemed that the monster was quieted; only the cracking and crunching of the bones of the animals he devoured could now be heard.

In the morning the Ases held council as to what was to be done; for the Wolf was slinking about, casting greedy looks at Asgard, as though he were devising how to break into the castles of the gods and carry off the spoil. They saw how gigantic he had grown, and knew that he daily increased in size and strength.

Heimdal pointed at Thor's hammer, and at Gungnir, the death-spear, in Odin's hand; but Allfather said gravely:

"The black blood of the monster may not soil the sacred courts of the gods. A chain must be made, so strong that it cannot be broken; then let him be bound with it, that his rage may be held in check."

The word was spoken, the work must be done. The Ases forged the chain Leuthing as quickly as they could, and took it to the Lyngwi island, where the Wolf, enticed by Tyr, followed them willingly.

The Wolf peacefully allowed himself to be bound, for he knew his own strength. When he was fully chained, he twisted and stretched himself, and the iron-ropes broke in pieces like weak thread.

A second chain, called Droma, much stronger than the first, was made, and he bore it for a moment; then he shook himself violently, and it fell clattering to the ground, broken to pieces.

The Ases stood round him silent and not knowing what to do, while Fenris increased his strength by devouring the food that had been thrown to him.

Wishfather now sent Skirnir, a young but wise and able servant of Freyer, to the Home of the Black-Elves, to get the Elves, who were versed in magic lore, and who lived in the bowels of the earth, to make fetters that should bind the Destroyer.

The underground people made a chain, small and slight as a silken thread, which they called Gleipnir. They said that it would grow stronger and stronger the more the prisoner strove to free himself from it.

Skirnir took the chain to the Ases. The All-Devourer resisted, and opened his mighty jaws threatening to swallow up all who tried to bind him; for he guessed that there was magic power concealed in the slight fetters.

Then brave Tyr came forward, petted and stroked the monster, and put his right hand into his jaws. Fenris thought this a sign that no evil was meant, so he allowed the slender chain to be bound around his neck and feet.

When this was done, he stretched himself violently, endeavouring to break his bonds, but they only became the stronger and cut into his skin and flesh. He had already bitten off Tyr's hand, and now he opened his blood-red jaws to seize the god himself and the other Ases too. But they feared the wild beast no longer; they thrust a sharp sword into his gaping mouth till the point penetrated the palate above and prevented him biting.

Then they fastened Gleipnir to two great rocks, that the Wolf might not get away. In vain the monster howled day and night while the blood ran down between his jaws and collected in the river Wan; he could not break his bonds.

Thus is crime, which threatens to corrupt the human race, bound by the apparently slight fetters of law, and as the power of the Wolf was broken by the sword, that of crime is kept under by the awards of justice. When a people no longer heeds the law, and throws aside all civic order, crime frees itself from its fetters, and the nation rushes to its ruin as surely as Gleipnir would be broken in the Twilight of the Gods, as surely as the All-Devourer would become freed from his chains and from the sword.

Tyr was called Tius by the Goths, Tio or Zio by the Anglo-Saxons, and the same by the Suevi, a tribe of whom, the Juthungen, lived beside the Lake of Constance. They were called Ziowari (servants of Zio), because they regarded this god as their guardian deity; the name of their chief town was Ziesburg (now Augsburg). The rune that stands for it, and is called after the god, is the sign of the sword. It bears the names of Tius, Tio, in Old High-German Zio, and besides these, is known as Eor, Erch, Erich, and in old Saxon Er, Eru, Heru or Cheru. These different appellations were all borne by the god, whose worship was so wide-spread.

Moreover the religion of the Suevi acknowledged a goddess Zisu, as is proved from the fragment of a Latin chronicle. She had a temple in Augsburg, and was of a warlike nature; she must therefore have been the female representative of the god Zio or Tyr. This god was the expression in ancient times of the impression that nature as a whole made upon the minds of those who were influenced by her. He was without form, and originally without a a name. When the Romans first knew the Germanic race he had already become a personality and was endowed with attributes, for they compared him with their own Mars, and therefore recognised him to be the god of war. Thus he had lost his original signification.

Tyr or Tius, meant brightness, glory, then the shining firmament, and was derived from the same root as the Hindu Djaus, the Greek Zeus, and the Roman Jupiter (Diu-piter, Dies-pater). Rays of sunlight and forked lightning both come from the sky, and were typified in arrows and deadly missiles. In the middle ages arrows were still called rays in German. Hence an arrow became the attribute and also the symbol of the omnipotent god of

heaven; in later times a sword took the place of the arrow as it was a stronger weapon in battle. This symbol remained to him in the rune and also in the groves which were dedicated to him. When his place was afterwards given to Wodan and Thor as the ruling gods of heaven, Tyr was looked upon as the god of battles, whose help must be entreated during the fight and whose rune of victory was scratched on the handles and blades of swords while ejaculating the name of the god.

Tyr was held in much less honour in the time of the skalds; he was then regarded as the son of Odin and the god of unnatural warfare that could never be appeased. Odin, the god of the mind, of martial courage and of poetic enthusiasm, had taken his place as the ideal of Kings and brave Jarls. Thor also, the god of the peasant, the benefactor of mankind, helped to force him into the background and gained some of the devotion Tyr had lost.

HERU OR CHERU, SAXNOT.

Nearly related to the warlike Tyr, perhaps identical with him, were Heru or Cheru and Saxnot. They were essentially German sword-gods, and were not known to the northern skalds. Their worship was wide-spread; for the Alanes, Quades, Getes and Markomanns paid divine honours to the sword, and even the Scythians, as Herodotus tells us, planted it in a high pyramidal heap of brush-wood, and called upon it as the symbol of the divinity. Many legends are still in existence about it, one of which we give as an example.

Cheru's sword was made in the mysterious smithy of the Dwarfs, whose artistic workmanship was celebrated among Ases and men. The sons of Iwaldi, who had made Odin's spear, and Sindri, who had forged Miölnir, had united their efforts in making the marvellous weapon on which the fate of kings and nations was to hang.

The zealous master-smiths worked busily within the earth, when Sökwabek was built under the flowing river, until at length the shining sword was completed, which Cheru the mighty god received.

This sword shone every morning on the high-place of the anctuary, sending forth its light afar when dawn arose, like a flame of fire; but one day its place was empty and the rosy light of morning only shone upon the altar from which the god had disappeared.

The priests and nobles sought the advice of the wise woman. This was the inscrutable answer they received.

"The Norns wandered on the ways of night; the moon had hidden his face; they laced the threads, strong and powerful, of gods and men, that none might break. One towards the east, the other towards the west, and one towards the south; the black thread towards the north. They spake to Cheru: 'Go, choose out the ruler, the lord of the earth; give him the two-edged sword to his own hurt.' He has it, he holds it in his hands; but yet Cheru the lord will bring it back after a time."

Startled at this dark oracle, the men begged for an explanation; but the maiden of the tower gave no reply. Meanwhile the story relates the course of events, and throws the only light that is given upon the riddle.

Vitellius, the Roman prefect of the Lower Rhine, was supping past midnight in his house at Cologne, for he liked the pleasures of the table better than all the glory and all the diadems in the world.

When he was told that a stranger, bearing important news from Germany, wanted to speak to him, he rose impatiently. He desired to get rid of him as soon as possible; but when he entered the anteroom, he found himself in the presence of a man of such distinguished appearance, that he could not treat him discourteously. He would have at once taken him for one of the Immortals, if his self-indulgent life had not long ago destroyed his faith in the religion of his ancestors.

The stranger gave him a sword of beautiful workmanship, and said:

"Take this weapon; keep it carefully and use it well, and it will bring thee glory and empire. All hail, Cæsar Augustus!"

The prefect examined the sword; when he looked up, the stranger was gone, and the guard had neither seen him come nor go. He returned to the supper-room and told what had happened. He drew the sword out of its sheath, and it was as though a flash of lightning passed through the room.

Immediately a voice exclaimed, but whether in the room or not, no one could say: "That is the sword of the divine Cæsar! All hail, Vitellius! All hail, Emperor!"

The guests at the supper-table joined in the cry and spread abroad the news; next morning the legions greeted Vitellius as Emperor. Messengers were despatched on horseback to the other provinces, and Fortune seemed to have chosen him as her favourite. His general conquered the army of his opponent, Rome opened her doors to him and the whole East acknowledged his sway.

"It was the sword of the divine Cæsar that made me master of the world," said the Emperor, as he seated himself at table to enjoy the delicacies which had been imported by land and water from distant countries. He ceased to care for the sword; he left it standing in a corner of the peristylium, where a Teutonic soldier of the body-guard found it and took it in exchange for his own clumsy old weapon.

The new possessor of the sword watched the conduct of the Emperor with displeasure, for Vitellius cared for nothing but the pleasures of eating and drinking; he paid no attention to the affairs of the Empire, or to the wants of the soldiers; he took no notice when far away in Asia brave Vespasian had been proclaimed Cæsar by his legions.

The German soldier left the Emperor's service and mixed himself with the idle populace. Meanwhile one misfortune after another befel the gluttonous Emperor. Provinces, generals, armies forsook him; the enemy's troops approached the capital; then Vitellius had recourse to the sword which had before brought him victory; but instead of it he found only an old and useless weapon.

Now all his courage forsook him; he wished to escape, and crept away to bury himself in a corner of the palace. The populace tore him from his hiding-place, dragged him through the streets, and when he reached the foot of the Capitol, the German soldier stabbed him to death with the sword of Cheru or of the divine Cæsar. In this manner was the prophecy of the wise woman fulfilled: "to his own hurt."

Afterwards the German soldier left Rome and went to Pannonia, where he re-entered the Roman service. He fought in many battles and was victorious in all, and soon became so famous that he was made centurion, and then tribune. When he grew old and was incapable of further service, he made a hole on the bank of the Danube, hid the good sword in it, and covered it up again with earth. Then he built himself a hut and lived there until his end. On his death-bed, he told the neighbours who had assembled round him, of his battles, and how he had got possession of the sword of Cheru; but he did not betray the place where he had hidden it, yet the saying that whoever should find the sword would become ruler of the world, remained current among the people from generation to generation.

Centuries came and went. The storm of the migration of races swept over the Roman empire; the Germanic races shared the spoil amongst them; the nomads of Asia, the wild Huns, made their way over from the East, like the waves of a sea, in order to have a share in the booty. Attila, or Etzel, raised his blood-besprinkled banner in the desire for land and military fame, but his efforts were fruitless for a long time.

As Attila was once riding with his troopers along the banks of the Danube, he busied himself with framing in his own mind gigantic plans of gaining for himself the empire of the world. He happened to look up and saw a peasant driving a lame cow and carrying a beautifully made sword under his arm. On being questioned, the man replied that his cow had hurt her foot against something sharp that was hidden in the grass, and that when he sought for the cause of the injury he found and dug up the sword.

The king desired that the sword should be brought to him, and drew it out of its sheath with joyful emotion; its bright blade shone fiery red in the evening light and all present stared at it in amazement.

But Attila, holding up the shining weapon in his strong hand, exclaimed :

"It is the sword of the war-god with which I shall conquer the world."

Having said this, he galloped away to the camp, and soon afterwards marched on to battles and victory. Whenever he drew the sword of the war-god the earth trembled from the east to the very west.

After his last campaign in Italy he married the beautiful Ildiko, daughter of the King of Burgundy whom he had slain. The youthful bride adorned herself unwillingly for the wedding she hated.

An old woman came to her secretly, and gave her the sword with which to revenge her father's death.

At length the king entered the bridal chamber in a state of intoxication and threw himself upon his couch. Ildiko now drew the weapon from under her dress and stabbed him to the heart with its sharp blade.

The rule of the Huns came to an end with the death of Attila, and the Germanic races chased these hordes back to the steppes whence they came; but tradition does not inform us whether these later deeds of war were done with the help of the miraculous sword. Yet it tells us of many strange things performed by means of it in the middle ages, and of how Duke Alba buried it in the earth after the battle of Mühlberg.

HEIMDAL (RIGER).

Once upon a time, when there was peace in the worlds, Riger arose and set out to visit his children of men, to see how they lived and what they did.

He walked along the green road, and arrived at last at a badly built house with a low roof. On the wooden bench beside the hearth were seated a man and his wife.

Ai and Edda (great-grandfather and great-grandmother) were their names, and they were very poorly clad. Riger addressed them kindly, seated himself between them, and ate with them of their coarse bran cakes, and their porridge in earthenware dishes.

The Ase remained in the cottage for three days and three nights, giving good counsel to them, and then went on from the sea-sand to the better ground for cultivation.

Nine moons after his departure a little boy was born to Ai and Edda, whose skin was of a dark colour and whose forehead was low. His parents called the lad Thrali. He grew and flourished, and soon learnt to use his strength. He tied up bundles with his muscular arms, and carried heavy weights upon his back all day long.

When he had grown to man's estate, he married a girl with black feet and sunburnt hands, called Thyr, who worked with the greatest diligence. From them are descended the race of Thralls.

Meanwhile Riger pursued his journey. He came to a roomy, well-built house in the middle of a cultivated field. There he found Afi and Amma (grandfather and grandmother) neatly dressed and working busily. The husband was making a loom, and the wife was spinning snowy linen thread on her wheel. A pot of good food was bubbling on the fire. Amma soon filled the plates, and at the same time gave her guest a cup of foaming beer as was the custom of the free-born farmer. Riger gave them much good advice regarding the management of house and land ; and after remaining with them for three days and three nights, he set out again along the road which ran through shady groves and across green meadows.

Nine moons passed, and then came a happy time, for a little boy was born to the great delight of his parents. He was called Karl (lad), and grew and flourished ; rosy were his cheeks, and bright and clear his eyes.

The boy soon learnt to drive the plough, to yoke the oxen and make carts in the same way as his father. In course of time he married Snör (cord), who was rich in keys and wore finely-woven dresses ; and he brought her home to his new house. Sons and daughters were born of this marriage ; all grew up active, merry, and free, and dwelt upon their own land.

Meanwhile Riger walked on through beautiful fields and blooming gardens up to the manor house on the top of a sloping hill. The door with its shining handle was not locked, so he entered the richly furnished hall. The floors were carpeted, and the father and mother were sitting on cushions, dressed in silken garments and playing with delicate toys.

Then the master of the house tried his bow, made arrows and whetted his sword, while his wife came out to watch him in a blue dress with a long train, and with a kerchief crossed over her white neck and shoulders.

Riger seated himself between them. He knew how to advise them for the glory and weal of their house.

Afterwards the lady spread the table with a beflowered linen cloth; she brought in well-cooked dishes of game and poultry, and filled the golden beakers and jugs with sparkling wine. They drank and talked till night-fall, and then Riger was shown his comfortable bed.

He remained with his hosts for three days and three nights, and then went away to continue his journey.

Nine moons passed, and a son was born in the manor house, fair-haired, with beautiful rosy cheeks and eyes like shining stars.

He was called Jarl; he grew and flourished, learnt to draw the sword, to throw the spear, to bend the bow, to carry the shield, to ride the horse, and to swim across the Sound. The boy learnt even more than this as he grew older, for Riger came to him out of the dark grove, and taught him to understand the runes, inspiring him at the same time to do deeds which should bring him and his house honour and glory.

Then Jarl went out to battle, conquered the enemy, and won for himself renown and booty, castles and land, rewarding his companions in arms generously with golden clasps and rings.

He became a great ruler, but still he felt sad and lonely in his luxurious hall. So he sent messengers to ask for the hand of Lady Erna, the slender-waisted. His offer was accepted, and the noble maiden entered his shining halls where the Earl received her with joy. They grew to love each other and lived together to a good old age.

Sons and daughters came of this marriage, and increased the number of the Jarls. The youngest son, Konur, understood the runes, both of the present and the future, and also the language

LOKI STEALS FREYA'S NECKLACE.

of birds. Besides this, he was a mighty warrior, and afterwards became the first King of Denmark. This is what the "Rigsmal," a poem of the Edda, teaches us of the beginning of class distinctions.

When Riger (or Heimdal) had finished his labours he mounted his horse, Gulltop (golden-mane), and rode home to Himinbiörg to fulfil his duty as watchman.

He drank sweet mead late each night, for all things in Asgard and without it were sunk in sleep. At midnight he once heard a noise of footsteps, but so faint was the sound that no ear but his could have heard it. It came from Folkwang, where Freya, the goddess of love and beauty, dwelt.

Heimdal cast a penetrating glance in the direction whence the sound came, and saw the sleeping goddess resting upon her couch. She was lying on her side, one arm resting upon her shining necklace, Brisingamen. Loki was standing beside her bed gazing covetously at the ornament. He seemed in doubt as to how he could get possession of it. He murmured magic spells, and lo! he grew visibly smaller and smaller. At last he became a tiny little creature, with bristles and a sharp set of teeth, a creature that thirsts for blood and attacks both gods and men; in the form of a flea he jumped upon the bed, and slipped beneath the sheets; he stung the sleeping goddess in the side so that she turned. The necklace was now free, and the cunning Ase, regaining his natural form, untied the ribbon that fastened it round her neck, and made off with it.

The faithful watchman on the heavenly tower was very wroth with the night-thief. He drew his sharp sword, and, as he had his seven-league boots on, came up with him in a few strides. He struck out at the robber, but his sword only went through a pillar of fire that towered up into the sky in which Loki's form had disappeared.

In a moment Heimdal rose in the shape of a cloud, from which such a torrent of rain descended that it threatened to extinguish the fire.

Loki immediately changed himself into a polar bear, that opened its mouth and drank up the rain. Before he could escape he was attacked by Heimdal as a still larger bear.

Loki fled from the deadly embrace in the form of a seal, but his flight was useless, for he was caught by another larger seal.

The two creatures fought furiously; they bit and scratched each other till the waters were stained with their blood. After a long and fierce struggle, Heimdal was victorious, and Loki slipped out of his torn and mangled seal's skin; but when Heimdal whirled his sword round his head, he begged for mercy and gave up the necklace to his opponent.

Heimdal stood leaning on his sword and holding Brisingamen in his left hand, rejoicing in his victory in spite of the pain his wounds caused him. But Iduna, Bragi's lovely wife, came to him and gave him an apple of eternal youth. As soon as he had tasted it, his wounds were healed and he ceased to suffer pain. He bade the goddess take the necklace back to Freya.

Then he returned to Himinbiörg, mounted his good horse Gulltop and rode down Iring's road, which men now call the Milky Way; immediately the black storm-clouds vanished and the shining stars lighted up the expanse of heaven in the same way that Brisingamen did Asgard's halls, until day came and called up gods and men to their work. For Heimdal is the same as Heimdellinger for Heimdäglinger, he who brought day to the home of the world. His name Riger shows that he was also related to the German Erich, Erk, Heru or Cheru, the sword-god, and consequently to Tyr or Zio. The Edda calls him the Sword-Ase, and makes him wander on the green ways of earth, as Iring did on the Milky Way, which was called after him. Certain roads bore the same name, such as those which ran through England from south to north, and the Irmin-streets in Germany that led to and from the Irmin-columns; thus Riger resembled the universal god, the giver of victory.

Riger's wanderings reminds us of Örwandil, whom Thor carried through the ice-streams Eliwagar. He was identical with the mythical hero Orendel, a son of King Eigel of Treves, whose travels and adventures on every sea have much resemblance to those of Odysseus. It is very doubtful whether these stories were known to the Teutons at the time of Tacitus, as this author mentions that the Hellenic hero had been in Germany, and had founded the town of Asciburgum (Ase-burg). It was rather to the poets of the middle ages that dark rumours of the Odyssee came.

Heimdal was born of nine mothers (the wave-maidens), whose names are taken from waves and cliffs; he was nursed and strengthened by Mother Earth, the cold sea and the rays of the sun; hence he appears as a god of heaven, raised aloft by the waves of the sea, which afterwards fall to the earth as fruitful rain or dew. This was his position in the natural myth. The skalds made him out to be the watchman of Asgard, to whom wa entrusted the care of Bifröst, the rainbow-bridge, that all attack of the giants might be prevented.

BRAGI AND HEIMDAL RECEIVING THE WARRIORS IN WALHALLA.

BRAGI AND IDUNA.

In the beginning the silence of death rested upon the immeasurable ocean, not a breath of wind stirred the air, not a wave rose on the surface of the deep; everything was motionless, dumb, without breath or life.

A vessel, the ship of the Dwarfs, crossed the silent waste of water. Bragi, the divine singer, was lying on the deck asleep, sunk in the dream of life; he was without spot or blemish, and his golden-stringed harp lay at his side. When the vessel glided over the threshold of Nain, the Dwarf of Death, the god awoke, touched the strings of his harp and sang a song that echoed throughout the nine worlds, describing the rapture of existence, the rage of battle and the charm of victory, and the joy and happiness of love. This song wakened dumb nature out of her trance.

Whether the god of poetry were the son of Odin or not, we cannot tell; the skalds do not inform us. But poetry cannot die, it always rises out of death to a new life and rejoices the hearts of both gods and men.

Bragi landed on the shore, singing his noble song about the awakening of nature and the blossoming of new life; and he wandered through the growing, budding woods as he sang. Then Iduna rose before him from amongst the grasses, flowers and foliage, the goddess of immortal youth, the youngest daughter of Iwaldi, the Dwarf, who hid life in the deep and afterwards sent it again to the upper world when the right time had come.

Iduna was beautiful in her crown of flowers and leaves; she was beautiful as the dawn. When the god saw her, his song of love became more glowing and intense. He stretched out his arms and she sank upon his breast, for the poet must needs marry youth and beauty.

After they were united, they went to the blessed ever-green heights of Asgard, where the Ases received them with joy. Then Iduna gave them to eat of the apple of ever-renewed youth.

When the gods and Einheriar had eaten their fill of the flesh of Sährimnir, Bragi touched the strings of his harp and sang the praises of the heroes. But this pleasant life in Asgard, and the married happiness of the divine poet, were once broken by a severe trial, as we shall presently see.

Odin, Hönir and Loki were travelling about the world together to see what were the joys and sorrows, works and labours of the dwellers upon earth. They went a long way, and at length came to a densely wooded mountain where there was nothing to eat. They could find no hospitable house in which to take shelter; could hear no friendly voice calling to them. The autumn wind was blowing the tops of the oaks and firs.

When they reached the valley, they saw a herd of cattle grazing in the meadow. They caught one of the animals and slaughtered it; they cut it up and prepared to cook it for their supper. The fire, kindled by Loki, blazed up, and they thought the beef would soon be cooked. But when they looked to see, it was still quite raw. This happened a second and a third time; the Ases were astonished and wondered what to do.

Suddenly they heard a voice above them saying that he who prevented the beef from cooking was sitting above them in a branch of the tree. On looking up they saw a gigantic eagle through the leaves of the oak, busily engaged in trying to put out the fire by flapping his wings. He promised to allow them to cook their supper if they would give him some of it. When they had agreed to do so, he flew down, fanned the fire, and very soon supper was ready.

They all sat down together, but the eagle ate so quickly that it seemed as though he would devour the whole bullock. Loki was dreadfully hungry, and getting into a rage, snatched up a stake and stabbed at the gigantic bird with it. The eagle flew up into the air when he felt the blow. The stake had fastened itself to the feathers of the bird and Loki's hands were glued to the other end.

The eagle flew so low that Loki's feet dragged along the ground and hit against any stones and stumps that might be in the way, while his arms felt as if they were dislocated. He shrieked and groaned and begged for mercy of the Storm-giant, who, as he well knew, was hidden under the eagle's dress.

"Very well," said the giant, "I will set thee free if thou wilt promise to bring me Iduna and her golden apples."

Loki swore to do so, and, as soon as he was set free, limped back to his companions. Under the circumstances the travellers determined to go home, and they must have been provided with seven-league boots, for they arrived at Asgard on the following day.

Beautiful Iduna was going about her household duties, dressed in green and wearing a garland of leaves, the crown of unfading youth. Bragi was away from home journeying as a minstrel. She collected her apples, which she usually gave the Ases at breakfast time.

At this moment Loki came up to her quickly, and looking round to see that no one was near, whispered:

"Gentle and lovely goddess, follow me quickly out of the castle gate, for I have discovered a strange tree covered with golden fruit like thine."

This was a request the goddess could not decline. She put some of her apples in a crystal dish and followed the traitor through Asgard, and on into the dark wood.

All at once the Storm-wind roared through the trees; and Thiassi, the giant in the eagle's dress, rushed up, caught the terrified goddess in his talons, and flew with her to dreary wintry Thrymheim, where spring flowers cannot bloom, not yet can youth survive.

Loki slunk back to Asgard, and quietly kept his secret about Iduna to himself. "The longer hence they notice it, the better," he cunningly thought to himself.

The Ases for a long time did not know that Iduna had been stolen; they thought she had gone away on a journey. But when days and weeks had passed their hair began to turn grey, the colour left their cheeks and their faces showed the folds and wrinkles of age. The goddesses, even Freya herself, discovered signs of approaching old age, when they looked at their faces in the mirror of a clear stream.

They all asked for Iduna and sought her high and low. The last time she was seen, she was walking with Loki. The cunning Ase was questioned; his lies did not help him; Thor threatened to break all his limbs, and raised his hammer for the purpose: then Loki confessed, and promised to bring back the giver of youth, if Freya would lend him her falcon-dress.

The request was granted, and he flew away at once to Thrymheim, the dwelling of the Storm-giant Thiassi.

The giant was at sea, and Iduna was sitting lonely and sad in an uncomfortable room, made of roughly hewn logs. Loki told her to be of good courage and changed her into a nut.

Then he flew over rocks and chasms with his light burden towards Asenheim.

Meanwhile the giant came home from his sea voyage. He had always hitherto begged his prisoner in vain to give him a slice of the apple of youth, that his horrible deformity might be transformed into the beauty of youth. As soon as he discovered Iduna's flight, he put on his eagle's dress and rushed after the fugitives with the speed of the storm.

The Ases watched the wild chase anxiously. They collected shavings and bits of wood before the fortress, and when the falcon had reached the shelter of the wall with his charge, they set fire to the wood, and the flames towered up into the air, singeing the wings of the pursuing eagle and bringing him to the ground.

Thiassi was then slain, but Thor threw his eyes up into the heavens where they shone henceforth as stars every night.

On his return, Bragi found his wife at home and heard from her all that had happened. He saw how Skadi, daughter of the Storm-giant, appeared in helmet and chain armour to avenge her father's death. And he afterwards told the whole story, ending with how Ögir, the god of the sea, had made expiation to the warlike maiden.

THE GIANT THIASSI STEALS IDUNA.

It is interesting to see how the genius of Odin's skalds united the god of poetry in marriage with the goddess of spring, the giver of renewed youth, and interwove the changes of the seasons into the myth. Bragi, who came out of the unknown distance, awoke mental life and also nature out of their trances; Iduna, who brought spring and youth into the world, became his wife. She gave the Ases the golden fruit of renewed youth, a fruit which was perhaps identical with the golden fruit that the Grecian hero Herakles carried away from the Hesperides.

In the same way as the autumn winds tear the leaves from the trees, the Storm-giant stole Iduna, and as the green meadows are covered with ice and snow in winter, so Iduna had to spend some time in the giant's uncomfortable house, while the gods themselves grew old and their hair turned grey.

Then Loki, probably the south wind, had to go and set Iduna free. The Storm-giant had gone on a voyage to the north, where his power lasted until the coming of spring. So the imprisoned spring was delivered from its bonds, and when the giant made his way into Asgard he was slain; *i.e.*, the storms of winter were confined within certain bounds.

ULLER.

Uller appears in the Edda as the cheery and sturdy god of winter, who, caring nothing for wind or snowstorm, used to go out on long journeys on his skates or snow-shoes.

Whenever he reached a lake or fiord which was not frozen, he transformed his shoes into a boat, and, making the winds and waves obey him, passed over to the other side.

Snow-shoes, as they are still worn in Norway and Iceland, are light shoes, very large and shaped like a boat turning up at the ends. With their help it is easy to slide quickly down hill, and they may have been the shoes alluded to in the stories of Uller; still, skates were also used at that time to glide over the frozen lakes. These shoes were also compared with a shield; thus the shield is called Uller's ship in several places.

When the god skated over the ice, he always carried with him his shield, deadly arrows, and bow made of the yew-tree. The pliable wood of the yew was the most suitable for making bows for use either in hunting or in war. Uller, therefore, lived in the palace Ydalir, the yew-vale.

As he protected plants and seeds from the severe attacks of the frosts of the north by covering the ground with a coating of snow, he was regarded as the benefactor of mortal men, and was called the friend of Baldur, the giver of every blessing and joy.

Once when out hunting, Uller saw beautiful Skadi, the bold huntress, of whom we shall have more to tell further on. He fell in love with her, and as she was by this time separated from her first husband, Niörder, she willingly consented to marry him. At the wedding the storms all played dance music in every tune, for the time when the day and night were of equal length in autumn was past, and winter, the happiest time for marriage, had begun.

Vulder with the Anglo-Saxons meant divine glory, or even God himself, and it seems that the Northern god Uller was thus characterised in heathen times. This was perhaps a consequence of the glory of the Northern winter night, which is often brilliantly lighted by the snow, the dazzling ice, and the Aurora borealis, the great Northern Light.

ULLER, THE BOWMAN.

Printed in Great Britain
by Amazon